# NATURAL RESOURCE MANAGEMENT AND INSTITUTIONAL CHANGE

The livelihoods of millions of people in developing countries are derived from the exploitation of natural resources, whether forests, rangelands, cultivated land or water. Recently there have been fundamental changes in the way resources and supporting services are managed and in the rights and responsibilities of resource users. New goals have been espoused and new institutions have been formed to pursue these goals.

*Natural Resource Management and Institutional Change* examines the effects of these recent changes, drawing on new evidence from a three-year programme of research in developing countries in Asia, Latin America and Africa. Government organisations have been privatised, decentralised or restructured while private sector organisations – both non-profit and commercial – have taken on increasingly important roles in resource management and service supply. Identifying the problems which stimulated calls for change, the authors analyse the strengths and weaknesses of different approaches to solving these problems.

This book provides an important and easily accessible point of reference for decision-makers and students alike, offering unique breadth of coverage across the natural resources sector and a range of different institutional types and approaches to resource management.

**Diana Carney** and **John Farrington** are Research Fellows in the Natural Resource Group at the Overseas Development Institute.

# ROUTLEDGE RESEARCH/ODI DEVELOPMENT POLICY STUDIES

1. NATURAL RESOURCE MANAGEMENT AND INSTITUTIONAL CHANGE
*Diana Carney and John Farrington*

2. DEVELOPMENT AS PROCESS
Concepts and Methods for Working with Complexity
*D. Mosse, J. Farrington and A. Rew*

Also available from Routledge:

IMF PROGRAMMES IN DEVELOPING COUNTRIES
*Tony Killick*
PB 0–415–13040–9

IMF LENDING TO DEVELOPING COUNTRIES
*Graham Bird*
PB 0–415–11700–3

MANAGING WATER AS AN ECONOMIC RESOURCE
*James Winpenny*
PB 0–415–10378–9

# NATURAL RESOURCE MANAGEMENT AND INSTITUTIONAL CHANGE

*Diana Carney and John Farrington*

London and New York

First published 1998
by Routledge
11 New Fetter Lane, London EC4P 4EE

Simultaneously published in the USA and Canada
by Routledge
29 West 35th Street, New York, NY 10001

Reprinted 1999

Typeset in Sabon by Routledge

Printed and bound in Great Britain by Biddles Ltd, Guildford and King's Lynn

*British Library Cataloguing in Publication Data*
A catalogue record for this book is available from the British Library

*Library of Congress Cataloging in Publication Data*
Carney, Diana.
Natural resource management and institutional change / Diana Carney and
John Farrington.
p. cm.
Includes bibliographical references and index
1. Natural resources – Management. 2. Natural resources – Developing
countries – Management. I. Farrington, John. II. Title.
HC59. 15. C37 1998
333.7 – dc21 97–44290
CIP

ISBN 0–415–18604–8

# CONTENTS

*List of illustrations*                                                vii
*Preface*                                                               ix
*Acronyms*                                                              x
*Acknowledgements*                                                      xi

**1 Introduction**                                                      1
*What types of reform?  2*
*What goals for reform?  5*
*What types of policy advice should we expect?  6*
*The structure of the book  6*

**2 The nature of the problems**                                        9
*Agricultural service provision  9*
*Forest resources  12*
*Water resources  15*
*Pastoral resources  17*

**3 Reforming the state**                                               21
*Types of change  21*
*Problems encountered in reforming the state  25*
*Conclusion  35*

**4 Non-state approaches**                                              37
*Common pool resource management groups  38*
*Farmer service groups  47*
*NGO approaches  55*
*Private commercial sector  62*
*Conclusion  72*

5 The interface                                                 74
   *Experience from Rajasthan 75*
   *Conclusion 84*

6 Summary                                                       88
   *Reforming the state 88*
   *Non-state alternatives 89*
   *Multi-agency partnerships 90*
   *Agricultural extension 92*
   *Irrigation and water resources 94*
   *Forestry 95*
   *Rangeland 96*

7 Conclusions                                                   97
   *What are the achievements to date? 97*
   *What are the main challenges for the future? 100*

   *Notes*                                                      105
   *Bibliography*                                               107
   *Index*                                                      116

# ILLUSTRATIONS

## Boxes

1 Definitions of the concepts underlying reform     3
2 Indigenous natural resource management in Totonicapan, Guatemala     13
3 Criteria for assessing robustness of local forest management institutions     44
4 Marketing groups in Uganda     48
5 Features of farmer groups usually associated with success     51
6 Increasing capacity for technology involvement: areas for attention     53
7 Establishing a farmer-to-farmer extension programme     58
8 Uncontroversial public sector tasks in a market economy     71

## Tables

1 Reforming the state: a summary     36
2 Non-state approaches: a summary     73
3 Characteristics of the main vehicles for process monitoring in inter-organisational collaboration in Rajasthan     82
4 Multi-agency approaches: a summary     87

# PREFACE

This book represents the culmination of a three year programme of research conducted within the Natural Resources Group at the Overseas Development Institute, London. The Rural Resources and Poverty Research Programme took place between 1993–6. It examined the changing role of the state and the changing rights and responsibilities of individual users of natural resources within the new climate of reform. It embraced the areas of agricultural services (notably agricultural research and extension), forestry, water resources management and pastoral resources management, and extended to three continents (Asia, Africa and Latin America). The Rural Resources and Poverty Research Programme comprised a number of different, but interlinked, activities including literature reviews, case studies, international networking and workshops. This book draws on all these sources and supplements them with the findings of further relevant research undertaken by ODI during this period, to present conclusions of broad relevance to developing countries.

# ACRONYMS

| | |
|---|---|
| ADP | Agricultural Development Project (World Bank, Rajasthan) |
| AKRSP | Aga Khan Rural Support Programme |
| ARF | Agricultural Research Fund (Rajasthan) |
| ASARECA | Association for Strengthening Agricultural Research in Eastern and Southern Africa |
| BRAC | Bangladesh Rural Advancement Committee |
| CDR | Complex, diverse and risk prone |
| CORAF | Conference des Responsables de la Recherche Agronomique Africaine (West Africa) |
| CPR | Common pool resource |
| DFID | Department for International Development |
| DoA | Department of Agriculture (Government of Rajasthan, India) |
| FF | Ford Foundation |
| GO | Government organisation |
| GoR | Government of Rajasthan |
| IPM | Integrated Pest Management |
| ITDG | Intermediate Technology Development Group |
| JFM | Joint Forest Management |
| KVK | Farm Science Centre (India) |
| MIRINEM | Ministry of Natural Resources, Energy and Mines (Costa Rica) |
| NABARD | National Bank for Agriculture and Rural Development (India) |
| NGO | Non-governmental organisation |
| ODA | Official Development Assistance |
| ODI | Overseas Development Institute |
| PEW | Para-extension worker |
| SYCOV | Malian Union of Cotton and Food Crop Producers |
| T&V | Training and Visit (extension system) |
| UNFA | Uganda National Farmers' Association |
| VLW | Village-level extension worker |
| WRDP | Water Resources Development Project (Rajasthan) |
| ZFU | Zimbabwe Farmers' Union |

# ACKNOWLEDGEMENTS

Various ODI research fellows and associates contributed to the work of the Rural Resources and Poverty Research Programme during 1993–6. Individual papers written for this programme are cited in the text; many have been published elsewhere. However, the various authors contributed in numerous other ways to this final thematic synthesis of the findings. We would like to thank, in particular, the following ODI research fellows: Roy Behnke, Jane Carter, Jonathan Cox, Mary Hobley, Michael Richards, Vanessa Scarborough, Gill Shepherd and Hugh Turral. In addition we would like to extend our thanks to the following ODI research associates: Ruth Alsop, Jonathan Davies, Elon Gilbert, Carol Kerven, Rajiv Khandelwal, Leila Mehta, Guillermo Navarro, Mahesh Pant, Kishore Shah and Alberto Vargas. Finally, a number of ODI staff assisted us with the manuscript at various times; particular thanks go to Caroline Dobbing and Alison Saxby.

The research programme was financed by the UK Department for International Development (DFID, formerly ODA, the Overseas Development Administration).

Much of the additional research on which this book draws was also funded by DFID, although the Rajasthan process monitoring work which forms the basis of a large part of Chapter 5 was financed by the Ford Foundation. We are grateful to funding agencies for their generous support over this period, but their funding should not be interpreted as endorsement of the views expressed here, which are those of the authors alone.

# 1

# INTRODUCTION

The past decade has seen a fundamental change in beliefs about the appropriate role of the state in developing countries. Rather than being the executor of a state-led process of development, the state must become the facilitator for a far more heterogeneous process in which a coalition of different actors and institutions is involved. This view has been reached through various changes in thinking. Between about 1950 and 1980 the primacy of the state was relatively unchallenged. During the 1980s there was a liberal backlash, with the 'New Political Economy' view that the state and those who worked within it were innately incapable of doing the job that they should have been doing. This view was based on the notion that public servants are motivated *only* by self-interest (summary presented in Toye, 1991) and given depth by more moderate thinkers who took the view that 'imperfect markets' are better than 'imperfect states' (arguments reviewed in Colclough, 1991).

Evidence to support this view was drawn from changes in developed countries. At the same time that socialist states were collapsing, and taking their economic paradigms down with them, many western countries were putting into place sophisticated and far-reaching privatisation programmes. The apparent (though not yet fully confirmed) success of these programmes strengthened the case for market-oriented reform in developing countries. An initial review of the evidence certainly seems to suggest that, over the past forty years or so, markets have done a fairly good job and developing country governments a fairly poor job. Compelling examples of this are provided by the 7 per cent per annum increase in agricultural production in some areas of China when private trade was legalised, and the wide range of goods available in the parallel market in countries where official trade is heavily regulated.

However, the problems associated with radical paring back of the state – especially in terms of equity, threats to the environment and the development of an excluded underclass – together with the fact that economic growth has been most pronounced in those countries of east Asia in which the state has not drawn back, have led in the 1990s to a growing consensus

1

that the state does have an important role to play in the development process. What is now proposed is the 'partnership' or 'pluralistic approach', which seeks to find ways in which the state can work together with NGOs, the private sector and people's organisations so that the 'comparative advantage' of each party can be exploited. The validity, indeed the necessity, of state involvement in the marketplace is once more recognised and efforts are being made to seek meaningful guidelines for the way in which all sides should operate (World Bank, 1997).

While the need for change can, arguably, be justified on theoretical grounds, it is more convincingly defended on empirical grounds.[1] Growth rates in developing countries have been unsatisfactory. Despite (and some would argue because of) significant inflows of aid and loans, standards of living remain unacceptably low for the majority of the population, especially in rural areas. Numerous instances of absurd waste and corruption and huge untapped potential have been identified by both outsiders and domestically-based critics. Even those in power have been forced to accept the desirability of change, as their coffers have run dry.

This book provides a synthesis of research conducted by ODI during the period 1993–6. The focus of the research was on the changing relationship between the state and the individual in the management of natural resources. It therefore examined both the processes of reform within government and the activities of civil society in taking over some of the functions previously performed by the state. In so doing it gathered early (but not comprehensive) evidence about reform in the various natural resource sub-sectors with which it was concerned (agricultural service provision, forestry, water resources and pastoralism). The overall goal of the research was to derive policy-relevant conclusions to inform the design of future changes.

This raises three important points. The first concerns the nature or scope of reform. The second concerns the goals of reform, and the third concerns the nature of the policy advice which can be drawn.

## What types of reform?

Deregulation, decentralisation and privatisation – the processes lying at the heart of the changes examined in this book – have all been defined in somewhat different ways. Our definitions for the purposes of this book are given in Box 1. The common element of each process is that the central government is shedding some of its responsibilities both for actual management of natural resources and for the supply of the services (such as agricultural research and extension) which enable people to make more productive use of those resources.

*Box 1* Definitions of the concepts underlying reform

PRIVATISZATION

Privatisation can be broadly defined as the process of change which involves the private sector taking responsibility for activities which were formerly controlled exclusively by the public sector. This may include the transfer of ownership of productive assets from the public to the private sector or may simply imply that 'space' is created in which the private sector can operate. 'Contracting out' of activities to the private sector would therefore be considered to be 'privatisation', but the development within the public sector of 'internal markets', 'cost recovery', or other modes of operation usually associated with the public sector would not.

DECENTRALISATION

Decentralisation is a process, a shift in the locus of power from the centre towards the periphery. Beyond this there is little consensus as to the meaning of the word. Some authors use it to refer to almost any move away from central government control, including privatisation. Here we take a narrower view. We focus on restructuring and changes in power relations *within* government. Decentralisation does not, however, imply that all power resides at the periphery. The centre still sets broad policy guidelines and goals and is responsible for coordination between decentralised units in addition to supplying certain key goods and services; getting the balance between the centre and the periphery right is one of the major challenges of decentralisation.

- Decentralisation within the law-making, legislative branch is referred to as *devolution*. This involves the creation or revitalisation of elected bodies at a lower level. It is unlikely to be successful unless some control over resources is ceded from the centre to these elected bodies.
- Decentralisation within the appointed bureaucracy, or executive branch, is known as *deconcentration*. This involves a shift in operational power away from the central ministry to sub-units outside the capital. It may coincide with a redefinition of the scope of a ministry but such a change is not, in itself, an example of deconcentration.

*continued* . . .

DEREGULATION

Deregulation and liberalisation are both rather general terms implying a loosening of government control and usually some form of privatisation coupled with increased competition. Paradoxically, deregulation, like privatisation, can actually imply a need for greater formal or explicit regulation of newly competitive markets. That is, while new players may be welcomed in, their activities, pricing strategies and general relations with their customers might be subject to significant outside (usually governmental or quasi-governmental) intervention.

(Carney, 1995b; Kirkpatrick, 1996)

What is sought here is not a reduction in public spending due to bankruptcy or indiscriminate cuts, but planned withdrawal from certain areas of activity. These areas of activity are either those which the state has shown itself to be very poor at managing or, more clinically, those which classical welfare economics tells us are better performed by the private sector (although advocates of reform would prefer these to be identical, it is clear that this is not always the case). The result which is sought is improved fulfilment of the ex-state responsibilities by new suppliers and, at the same time, improved performance by the state in its residual areas of activity. This is a persuasive argument: if resources are focused on a narrower range of activities then, other things being equal, it seems safe to assume that performance should at least not deteriorate. In the core argument, well-planned and well-executed withdrawal reduces costs, improves government performance and – some would argue – stimulates growth in that it provides the right incentives for further private sector activity. A knock-on effect of this, according to some, is that jobs and income created by the private sector lead to a more equitable diffusion of the benefits of growth to more people (OECD/DAC, 1995). However, poorly planned withdrawal clearly does not bring such advantages. It may deter the private sector from engaging in new economic activity (because of residual state interventionism, inconstant state behaviour, or inappropriate supporting regulations and infrastructure) and may pose major threats particularly to the livelihoods of the poorest and to the environment.

## What goals for reform?

The concepts which now dominate the debate surrounding the role of the state in rural development are effectiveness, efficiency and accountability; all three are related, as will be outlined below. It is important at the outset to clarify the way in which each is used in this book. *Effectiveness* refers to the ability to meet goals, objectives or needs – in this context, primarily those of the rural population. *Efficiency* refers to the way in which goals are met, and implies that this is done at as low a cost as is possible without compromising effectiveness. *Accountability* is institutionalised responsiveness to those who are affected by one's actions. Since the responsiveness is formalised, measures must exist to enforce it (these are usually electoral or financial).

Thus, effectiveness is a precondition for efficiency and accountability can enhance both. In achieving efficiency individuals or organisations should be identifying critical tasks in conjunction with users, then striving to find the most cost effective way of performing these tasks and cutting out superfluous tasks. This implies a culture of continuous learning. Institutionalising such a culture represents a formidable challenge.

It is important at the outset to clarify one theoretical issue. Efficiency, as defined above, does not always coincide exactly with *economic efficiency* as traditionally understood. In general, economic or Pareto efficiency is achieved when nobody can be made better off without somebody else being made worse off. In the absence of market failure, markets are considered to be the best allocative mechanism for achieving this. However, if governments wish to service the needs of those sections of the population with inadequate purchasing power which are not served by markets, they will need to intervene in pure market allocations. This does not imply that they entirely sweep aside such allocations but, rather, that they supplement them in particular areas (for example, providing funds to which poorer users can apply under certain conditions in order to purchase goods or services available in the market place). According to the definition above, they can do so and still be efficient (so long as they adopt the least cost method of doing so), which would not necessarily be true given a narrowly economic definition of efficiency. Such 'social efficiency' can, therefore, be congruent with a strategy of poverty reduction, an objective which now takes centre stage in many donors' hierarchies of goals (including those of the UK Department for International Development (DFID)). Such efficiency might thus be thought of as '*social efficiency*'. This is not, however, an easy concept with which to work since it has no natural boundaries. To the extent that it involves meeting people's needs, there must always be an additional decision taken about how far the public sector should go, given scarce resources and inevitable tradeoffs between provision in different natural resource subsectors and even different sectors (for example, tradeoffs between spending on education, health, industry and so on). Furthermore, if the public sector goes too far down this route it will be in danger of regaining ground from which reform had intended to displace it.

It is also important to note that increases in efficiency and effectiveness will have to be pursued in the context of underlying national objectives. Issues of inter-generational equity and concern about preserving the overall resource base are particularly pertinent;[2] these may conflict with demands made by current users of natural resources. States must recognise inherent conflicts such as these and take measures to adjudicate in a rational manner.

## What types of policy advice should we expect?

While it is intuitively appealing, the concept of 'social efficiency' is relatively difficult to handle because of the degree of flexibility that it entails. The underlying language of reform is one of responsiveness to local needs, facilitation of local initiatives and respect or support for existing management practices (where these function). This in turn raises a plethora of new issues relating to information management and the nature of the policy guidance that can be provided. The purpose of policy advice is to provide the foundation for concrete actions, yet the type of policy advice that flows from the new quest for effectiveness is seldom definitive. It consists instead of 'best practices' which have been identified but will probably need to be adapted to a given situation. While this type of advice is more realistic than are sweeping prescriptions, it has its own complications. Solutions are viewed as situation-specific, which implies that government staff, donors and other decision-takers must be able to interpret the environments in which they work and to make plans accordingly. This entails levels of skill and analytical availability which many current staff may not have or, certainly, may be unused to exercising. Hence, there is a need to stress human resource development as an integral part of the reform process.

Nevertheless, given limited time and resources for exploring the full details of every situation, we must look for general guidelines and focus on putting in place structures and decision-support systems which will facilitate effective management of resources with limited on-going external input. This explains the level of generality which is sometimes necessary when drawing conclusions and making policy guidelines.

## The structure of the book

This introduction outlines the scope and complexity of reform in the management of natural resources and supporting services. Chapter 2 elaborates on a variety of situations which demonstrate the need for reform in the various natural resource sub-sectors with which we are concerned (agricultural service provision, forestry, water resources and pastoralism). Our focus overall is on meeting the 'productivity enhancing' needs of the rural poor. We do not go into all the many justifications for doing so. Let it suffice to say that, given the continuing prevalence of poverty in rural areas, as well as the intimate relationship between agricultural production and food security, this is an objective which should be at the top of most political agendas in democratically controlled developing countries (Garcia, 1994). Only if it can be achieved will the massive projected increases in demand for food over the forthcoming decades represent an opportunity rather than a threat to poor, rural people.

Chapter 3 goes into detail about the process of reform within government, looking at both the problems this entails and some of the successes which have been achieved. As with all reform, there are many problems which are not revealed until the process is underway. What we produce here is, therefore, an early attempt to bring together some lessons from reform so as to facilitate the learning process for reforms of the future. Once again, one of the critical pieces of advice is 'know your environment', including both the overt and the covert institutions which influence change. An 'actor-centred' approach is useful if the underlying reasons for the failure of much reform are to be exposed. Frequently reform is designed to, or incidentally happens to, emasculate the power of a few key individuals. It is only by gaining an understanding of who exactly is under threat and how they might derail or pervert the course of reform that the dangers which this brings can be avoided. Time spent courting political support for change and building in safeguards against elite capture, particularly when reforms are in their delicate early months, is rarely wasted.

Chapter 4 reviews the situation from the 'bottom up'. A critical distinction is made here between the management of natural resources and the provision of supporting services. In the former case, the emphasis is on management by the users of the resources themselves, often through the medium of groups and with the assistance of supporting NGOs. In the latter case, while groups (for example farmers' organisations) may play a role, it is often a question of third parties (neither users nor the public sector) taking on new responsibilities and intermediating between the two sides. Non-membership NGOs play a part here as does the private commercial sector.

In Chapter 5 we focus on the interface between the different suppliers and managers of resources and supporting services, drawing particularly on a case study from Rajasthan. It is the firm conclusion of this research programme that only joint (i.e., multi-agency) approaches are likely to succeed in overcoming the problems which have for so long held back the process of rural development. We are not 'public sector pessimists'. We therefore reject the view that since market failure is better than government failure, government activities should be reduced to a minimum across the board and that, at best, government should be a broker to adjudicate between different, potentially conflicting, interests (Colclough, 1991).[3]

Neither are we 'public sector optimists' (who focus on the imperfections of markets as a reason to implicate the government in all areas of activity). We therefore believe that all parties have a role to play and that the nature of that role is likely to evolve over time.

Finally, Chapter 6 presents a summary of the research and draws out some specific recommendations for policy makers. It provides an overview of achievements to date and takes note of some of the major challenges which are expected to be faced over the coming years.

# 2

# THE NATURE OF THE PROBLEMS

This chapter examines the nature of the problems which have generated calls for reform in each of the four sub-sectors with which we are concerned: agricultural service provision (focusing on extension), forest resources, water resources and pastoral resources.

Although there is a good deal of commonality between the problems in all sub-sectors – for example, high recurrent costs and lack of participation by users in management – there are also some concerns which are specific to certain sub-sectors. For example, problems of clashes between commercial companies and local users are most often encountered in the area of forestry because of the high value of forest resources. Cost overruns, on the other hand, are usually associated with water resources or, to be more precise, the construction and maintenance of large scale irrigation schemes. We therefore review the key problems of each sub-sector in turn. This illustrates the pervasiveness of certain problems but also the fact that solutions must always be context- and resource-specific.

## Agricultural service provision

In the past, agricultural service provision tended to be monopolised by the public sector. Government monopolies were insulated from public pressure, and research and extension organisations isolated from their clients (Antholt, 1994; Sims and Leonard, 1990; Merrill-Sands and Collion, 1993). Extension workers 'transferred' inappropriate technologies and inputs failed to arrive in time to fit in with the agricultural calendar (Okali *et al.*, 1994; Röling, 1990; Sims and Leonard, 1990). Marketing boards were a particular problem, frequently being wasteful, corrupt and unresponsive to changes in supply. They neither provided adequate price incentives to farmers nor, in some cases, viable purchasing structures (World Bank, 1994).

However, the problem was not simply the potential for abuse inherent in the monopoly situation, but also a fundamental lack of concern on behalf of governments, individuals within governments and donors for rural producers. At best they were treated as ignorant and in need of modernisation, and at worst they were actively exploited by urban elites (Schiff and Valdés, 1992; Bates, 1981; Jaeger, 1992). Consequently, public sector institutions in many countries made insufficient efforts to identify and service their needs, let alone incorporate them into their own criteria for staff evaluation.

### Agricultural extension

During the past two decades, extension has been dominated by the Training and Visit (T&V) approach. This has been the philosophy behind some US $5 billion of investment by the World Bank alone. T&V is characterised by:

- a single line of command;
- a stripping away of services not integral to the provision of advice;
- a focus on 'contact farmers' (more recently, 'contact groups');
- time-scheduled activities;
- regular training and refresher courses;
- an aim to develop close linkages with research.

These features remedied several of the problems of earlier approaches to extension and addressed some of the problems associated with the management of large, national extension systems. In particular, structures were simplified, targets well-defined and opportunities for corruption reduced as extension services withdrew from input supply activities. Nevertheless, T&V has a number of significant weaknesses:

- the messages it promotes are often too rigid for the diversity of farming systems on the ground, being fixed in both timing and content;
- the number of relevant messages for difficult environments is limited;
- the needs of women farmers have generally been neglected;
- achievements are often measured in terms of quantity (e.g., number of visits, training sessions, etc.) rather than quality (e.g., rate of satisfaction of client farmers).
- it does not provide for multi-agency approaches involving private commercial or non-profit organisations, or existing farmers' organisations;
- the 'contact farmer' mechanism rarely works as well as intended, and inadequate group formation skills among extensionists mean that 'contact farmers' have not always been successfully replaced by 'contact groups'.

In Bangladesh, the public sector extension system has performed poorly against criteria of coherence, performance and sustainability. The T&V system provided a structure for ensuring that field-level extension workers received regular 'messages' to transmit to farmers, received training and backup as necessary, and made regular visits to contact farmers. However, the fact that formal performance criteria were met did not imply that the system was effective. Messages were derived from 'ideal' station environments without taking into account the state of the input, credit and output markets. They were therefore tailored neither to farmers' needs (in terms of location specificity, range of crops covered and timing of agricultural activities) nor their capacities (Chowdhury and Gilbert, 1996).

One reason why the T&V system was unable to adapt and become more responsive to farmers' needs was a lack of resources for operational activities. By 1989, 95 per cent of the budget of the Department of Agricultural Extension was dedicated to the payment of the salaries and allowances of the 24,000 staff. This problem was not unique to Bangladesh. In 1990–1, salaries accounted for 88 per cent of total spending on extension in the Indian State of Tamil Nadu and only $1 per annum per extension agent was allocated for 'materials and supplies' (Antholt, 1994). A recent World Bank evaluation of completed extension projects, all of which were located within and designed to assist the public sector, shows that 90 per cent have experienced recurrent-cost funding problems and 70 per cent are probably not sustainable (World Bank, 1994). It is unfortunate, then, that T&V does not recognise the potential roles of private commercial or non-profit (i.e., NGO) organisations as complementary sources of technical information or as a means of mobilising farmers, for this means that it does not recognise their potential role in cost-sharing.

Practice has not lived up to rhetoric in terms of improving linkages between research and extension under the T&V system. Feedback from farmers to researchers has remained weak, and researchers have been reluctant to use the problems identified by farmers as a basis for research prioritisation. Partly for this last reason, T&V has been unable to respond easily to pressures towards participation and good governance.

Finally, changes in the likely nature of future innovation raise broader questions about the role of the public sector in the provision of extension advice. The Green Revolution was based on technologies which were widely adopted (especially, but not exclusively, by richer farmers in higher potential areas) and which had been developed and disseminated by the public sector. It is likely that future gains in agricultural productivity through technological innovation will have to be more incremental, knowledge-based, locally-specific and directly geared towards overcoming specific farmer constraints. This is particularly true for resource-poor farmers operating in environments which cannot be unified through irrigation and purchased inputs, which are remote from markets and political and urban centres, and in which the natural resource base is fragile. It is precisely in areas such as these that the state tends to be under-represented. The need for locally-specific technological innovation means that, if agricultural research and extension organisations are to be effective, their agendas and outputs will have to be more demand-led than they were in the past (Scarborough, 1996; Antholt, 1994). It remains questionable whether, even with greater access to resources, the state can manage such diversity, though decentralisation of management structures would be one important step in the right direction (provided that safeguards against domination by local elites are built in) (Carney, 1995b).

## Forest resources

Control of forest land in developing countries has, over the course of the twentieth century, tended to be vested in the state. What this means in practice is that the state has had the power to allocate rights of use to local people or private companies as it sees fit. It rarely means that the state has effectively managed the forest land itself. In most settings, it simply does not have the resources to do so. The vastness of the forest areas nominally controlled by the state makes it difficult for it even to monitor those agencies to which rights of use have been granted, and to prevent public sector staff in isolated locations from abusing their positions. In India, forest departments control 22 per cent of national territory (Agarwal and Narain, 1989). In Indonesia, the Forest Department administers a massive 74 per cent of the nation's landmass, while the equivalent figure for Thailand is 40 per cent (Colchester, 1994).

State ownership of forest land, coupled with stronger efforts by the state to control the use of this land, has had two main effects. First, in combination with rising population pressure and sometimes internal factionalism within communities, it has disenfranchised traditional users of resources and often led to the destruction of existing regimes for forest management (see Box 2). Such local management regimes have been unable to survive the superimposition of economically motivated or market-driven management practices which have been introduced by the state in its nominal quest for efficiency. Traditional societies based on an economic system which emphasises reciprocity have had to grapple with a new monetisation of their values and the imposition of an economic system based on individual material gain. Although not all have failed, there are plenty of examples of those which have and where, as a consequence, traditional management regimes have been rejected (Richards, 1997a).

*Box 2* Indigenous natural resource management in Totonicapan, Guatemala

The Maya of Totonicapan have managed dense forest areas known as *parcialidades* for centuries. Veblen (1978) reported the following protective mechanisms or institutions:

- community-elected forest guards to keep out outsiders and check forest product use;
- regular community meetings to discuss problems and possible solutions;
- permission required for anyone wanting to cut a tree, with those who disobeyed being punished (for example, the right to graze sheep was suspended or a fine was imposed);
- a tradition that each family could fell two trees per year for household requirements;
- a selective felling system, in which the best reproductive trees were left;
- belief that trees have spirits and that their place and function in the universal order is as important as mankind.

However, over the last 10–15 years, several community protection mechanisms have ceased to function or have become less effective due to the emergence of a new state regulatory code, and other exogenous and endogenous pressures such as the demise of traditional Mayan cultural beliefs.

State authorities attempted to introduce a licensing system to regulate tree felling, but this proved costly and time-consuming and many communities became alienated. At the same time, armed gangs systematically stripped white pine of its bark for use in leather curing, mainly in night-time raids; a count in 1991 revealed that in six communities, almost 6,000 trees were stripped. Police have been uncooperative, sometimes even supporting the bark strippers by arresting local vigilante groups as 'guerrillas'. Complaints to higher authorities proved fruitless in the face of suspected bribery. Participation in reforestation schemes has been abandoned, as local people see no point in joining schemes when the potential benefits are likely to be seized by the bark strippers. Community institutions have been further weakened by intra- and inter-community conflicts over land rights (especially where land tenure documentation has been unclear).

(Based on Utting, 1993)

Second, state intervention has not tended to preserve the forest land or to develop it in a sustainable fashion. The irony is that local management of forest resources was initially rejected for fear of the 'open access' which it appeared to entail. The state stepped in to reaffirm its control and to take on the position of guardian of the forests. However, in so doing it has hastened the transformation of forests into open access resources (where existing local management regimes have been destroyed and not replaced) or has tacitly (or actively) encouraged unsustainable degrees of exploitation. A number of forest departments, particularly in Latin America, were established with the explicit target of supporting industrial development and maximising growth. They have therefore granted logging concessions to exploitative domestic and international commercial companies, the immediate attraction of generating funds triumphing over the imperative of maintaining forest resources for the future. The state has also allowed or encouraged relatively uncontrolled settlement of colonists on forest land.

Another response has been to privatise forest land completely (as in the Brazilian Amazonia). One hope has been that this might reduce the corruption that appears to have been endemic in many forest departments the world over. Where individuals or companies have been granted large areas to manage, existing users, many of whom are landless, are disenfranchised (with adverse equity implications and sometimes violent outcomes). Even the economic returns to such strategies have been questionable, because of the low productivity of most of the cattle ranching systems which tend to replace forests following privatisation and the initial felling of old-growth timber. While areas which lie close to the household can be effectively handed over to individuals or families, which might appear to be a more equitable approach, larger areas cannot be treated in the same way. Boundaries are generally too long and permeable for individuals to police them, so the result of such initiatives is to hasten the degeneration into an open access situation.

At the time that most forest departments came into being (in the early twentieth century), conservation objectives and concerns about local input into decision making had little influence. New concerns for participatory development and conservation of biodiversity have now come to the fore in many countries, but traditional forest departments have found it hard to adapt to the new multiple-objective demands (Richards, 1995). Distant state authorities are rarely able to allocate rights of use in an effective manner, and the consequence is that resources are overexploited and the poorest tend to lose out. Recognition of this and the fact that effective management of forest resources cannot take place without the cooperation of users has militated towards the joint forest management approach in India, and also elsewhere in south Asia and parts of Latin America, for example, Mexico (Richards *et al.*, 1996). This is not without its own difficulties (see Chapter 3).

In addition to increasing pressures for democratisation and concerns about sustainability, internationally imposed structural adjustment pressures have played a part in reform of the forest sector. There has been an imperative to scale back the efforts and the size of forest departments and to define a new role for the state as it attempts to meet the challenge of developing community-based forest management. Amongst other things, this has led to a move away from resource creation projects (centrally planned plantations) towards institutional reform programmes. Decentralisation of forest institutions, and the challenge of how to effectively promote community-based management, was therefore the focus of two of the field-based pieces of research conducted by the Overseas Development Institute (ODI) during 1993–6 in India (Hobley and Shah, 1996a) and in Central America (Richards *et al.*, 1996).

## Water resources

There are a variety of different problems, and hence sources of pressure for reform, in the water resources sub-sector. It is notable that while some of these (such as the poor maintenance of irrigation systems) have been relatively broadly addressed (with varying degrees of success) others, such as the increasing conflict between urban and rural water users or the lack of specification of water rights at a local level, have thus far been largely ignored, or addressed only in an *ad hoc* manner.

The major concern in the water area has been to reduce the amount of public money which has been spent on financing and operating large scale irrigation systems (many of which have been of dubious productivity). For example, since 1940, 80 per cent of Mexico's agricultural investment has been spent on irrigation. Similarly, over recent decades, China, Pakistan and Indonesia have all dedicated approximately 50 per cent of their agricultural investment to irrigation. A significant portion of the funding for irrigation schemes has come from donors, most of which now view large-scale infrastructural investments with some suspicion. There have been too many examples of ill-conceived schemes, inability or reluctance to maintain existing schemes and misuse of donor funds for irrigation schemes to remain attractive investments.

Despite the initial success of irrigation in supporting the Green Revolution in Asia, irrigation schemes have often underperformed in economic terms, and field research has highlighted substantial shortcomings in management (operation and maintenance), equity, cost recovery and agricultural productivity. Turral (1995a) lists reasons for this as:

- unrealistic productivity projections at appraisal;
- capital cost overruns;
- substandard construction or design;

- poor system management and service provision;
- poor understanding of farmer priorities and inadequate markets for produce.

In addition, as in the forest resources sub-sector, the increase in externally constructed and managed irrigation systems has been shown to lead to the neglect of pre-existing farmer-managed irrigation systems. Investments in physical infrastructure have been made at the expense of investments in 'social infrastructure', with a detrimental effect on overall system performance (Ostrom, 1992).

Another problem has been lack of clarity about responsibilities for irrigation management and water policy more broadly. In Nepal, for example, at least four separate ministries or bodies are involved in developing irrigation facilities and drinking water provision (Pant and Cox, 1996). Broadly speaking, however, irrigation departments have tended to dominate water policy-making in developing countries, and within these departments the construction lobby has defined the agenda. One reason for this has clearly been the significant opportunities for patronage and rent seeking that this affords to decision-makers, both domestic and international.

At the same time that these concerns were coming to the fore (in the 1980s), it was becoming clear that private management of certain water resources represented a viable option. In Bangladesh, small-scale water markets appeared to be developing and functioning in a much more effective way than any of the larger irrigation schemes. Various traditional water management regimes were identified in which use rights were highly, but successfully, regulated. These various influences gave birth to the move in many countries towards 'irrigation management transfer', which has involved various degrees of withdrawal by the state in favour of management of existing schemes by user groups or intermediary private companies. A whole new area of policy debate, focusing on questions of the responsibilities of the different parties and the implications of change for equity and system performance, has since emerged.

The focus in the water area, as in the other sub-sectors, has therefore been on institutional reform. This is despite the fact that improvement in water management, more so than the management of the other resources with which we have been concerned, appears to depend on a combination of improved institutional and technological approaches. The main cause of the need for technical innovation lies in the unprecedented increase in urban demand for water and sanitation which is already being witnessed, and which is expected to accelerate over the forthcoming decades.

The full implications of the conflicts between urban and rural water users have yet to be addressed, and farmers continue to benefit from privileged or under-priced access to water in many countries. For example, in Tamil Nadu the zero tariff on electricity used for agricultural water supply is considered

16

to contribute substantially to the overuse of groundwater (Turral, 1995c). In other places, however, water rights are being summarily taken away from farmers and allocated to urban or industrial users, without adequate analysis of the underlying efficiency of different types of water use. Though absolute scarcity of water is expected to be a growing problem in a number of countries over the coming decades (resulting in higher food prices), more rational use of existing resources is expected to go some way to solving the problems of competing demands for water.

## Pastoral resources

Nearly all African rangelands are state property, but this legal designation rarely has much practical significance. In reality, a number of different types of management regime can be identified, each of which has associated problems. Indeed in the area of pastoral resources (perhaps more than in the other areas with which this research has been concerned), general lessons for effective management have been slow to emerge. Part of the problem lies in poorly developed understanding of existing practice.

For example, Cox and Behnke (1995) point out that our understanding of tenure arrangements in Namibia is 'inadequate'. They blame this partly on Namibia's colonial history and the fact that most descriptions of tenure arrangements have been passed from author to author as 'received wisdom', creating the general impression that communal tenure is monolithic and static. Drawing on evidence from Ethiopia, Kerven and Cox (1996) argue that traditional institutions and their interpretations are neither easily observable nor static. Territorial boundaries between clans are not necessarily fixed in either physical or social terms, and change tends to take place organically in ways which are far from transparent. Both sets of authors argue for more work to be put into documenting the intricacies of current management systems.

The following appear to be the typical 'management' scenarios for rangelands. The problems of each are discussed in turn.

### Management void

When rangelands are truly 'open access' rather than indigenously managed, they are often overexploited. If no management regime is in place, economic rationality directs users to exploit rather than conserve with resultant 'tragedy of the commons' effects (Hardin, 1968). Furthermore, it is often richer people who benefit at the expense of 'average' or poorer users. In Namibia, lack of clarity about the legal status of land has led to semi-legal fencing of land by the elite (thereby increasing the incidence of damaging over-stocking in the areas which have not been fenced). In this case it is only the elite who can afford the fencing materials and who have the tacit support of the government, which still views privatisation of land as the route to progress and advancement. In Ethiopia, on the other hand, there have been periods during which government authority has been entirely absent. During the period of transition from the Dergue government, various ethnic groups lost territory and livestock as the state's authority collapsed and its ability to assure security disappeared (Kerven and Cox, 1996).

### Semi-legal indigenous management

In the absence of effective state management of pastoral areas, pre-existing indigenous management regimes have frequently persisted. These can be effective, even where they are not equitable. Perhaps their main advantage lies in their flexibility and the fact that they can evolve to embrace change as it occurs. Behnke (1995) argues that African tenure has shown itself not only to be adaptable but also to be able to reproduce many of the supposed benefits of individual titling of land. One reason for this is that indigenous management regimes are rarely as communal as is assumed by those unfamiliar with their details. For this reason, indigenous pastoral management may be able to limit overexploitation.

However, indigenous management poses other problems. The assertion of rights through armed conflict, and the creation of insecure inter-tribal buffer zones in which resources are under-utilised are two particular problems. In Ethiopia, conflicts between the Boran and the Gabbra people have led to a decline in the use of grazing reserves in lowland areas and an associated move towards sedentary agriculture in undisputed upland areas. Indigenous management systems can also prove inadequate to the tasks which they face (especially when they are required to adapt to some level of external intervention). Again in Ethiopia, elders in the Borana Plateau stated that solving problems of degradation and overcrowding was too great a challenge for the local management committee (Kerven and Cox, 1996). Likewise, the pressure towards enclosure in northern Namibia was too great for traditional management structures to resist, even though local people were vehemently opposed to the fencing activity.

Finally, problems with indigenous management occur when current managers are challenged, whether by the state or other would-be managers. Since they do not have recourse to legal enforcement mechanisms, users of resources within indigenous systems tend to lose out in the face of conflicting claims. However, translating customary rights into written law is such an immense task that it can easily overload the legal system (Behnke, 1995). In many cases there also exists the problem of determining how 'genuine' customary law is. In Namibia, for example, many features of today's customary law date back no further than fifty years, to a period when tribal 'custom' was commonly subverted by the South African regime.

## Conflicting state/indigenous management

Probably the most problematic management scenario is one in which multiple tenure systems coexist. This is frequently the case when new 'formal' management systems are superimposed upon existing indigenous systems. This was a common occurrence when prevailing beliefs held that 'modernisation' through formal titling was required for pastoralists to be able to access markets and obtain the credit which would allow them to prosper. Mixed management systems cause legal ambiguity and increase the costs of maintaining property rights by forcing proprietors to defend themselves in multiple forums. Confusion can also lead to breakdown of both systems as was the case when the socialist Dergue government in Ethiopia tried to replace traditional social mechanisms for allocating and managing natural resources with a state-managed system based upon elected local officials and committees of pastoral associations. The association of grazing land with these 'state imposed' bodies meant that it was subject to looting which would not have been the case had the traditional management regime still been in place.

## Project approach

The 'project' approach to rangeland management constituted an integral part of various African governments' rangeland modernisation programmes during the 1980s. Depending upon the intensity of effort, the approach might aim totally to replace indigenous management systems with externally imposed regimes or, perhaps more common in outcome if not intention, to draw on aspects of both indigenous and 'modern' management. Where the intention is to totally replace indigenous management, the mechanism is usually wholesale privatisation of access to land. During the 1980s, the state in many African countries tried to update traditional tenure systems through projects based on the notion of 'ranching' (inherited from Australia and the USA). The design of these projects was founded on a belief that grazing lands were over-stocked and that one of the roles of government should be to facilitate marketing, so as to capture benefits in terms of increased incomes for herders, increased supply of livestock products to meet urban demand, and reduced livestock population pressure (hence rangeland regeneration).

However, Behnke and Kerven (1994) argue that such projects were misconceived. They failed to take into account the full range of benefits generated by pastoral systems and the flexibility which is inherent in them. They also neglected to take into account the fact that if units which are economically viable on a year-round basis are to be created, then they must cover so large an area that many pastoralists would, of necessity, be dispossessed. Since pastoralists are able to follow highly erratic rainfall patterns,

19

they can maintain higher livestock densities and hence higher productivity per hectare than their ranching counterparts. Kerven and Cox (1996) note that while range managers increase immediate productivity, they do so at the expense of long-term productivity: 'the communally-managed rangelands of the Borana [Ethiopia] exhibit few signs of degradation compared with the adjacent uplands which have been carved out into privately managed fields.'

Projects which have aimed to supplement indigenous management regimes by increasing the services available to pastoralists have often encouraged unsustainable livestock densities. Again in Ethiopia, projects which have focused on developing infrastructure and access to water (through ponds) and veterinary services have led to larger, less stable herds and a rise in human population (Kerven and Cox, 1996). Alternatively, those designing the projects have assumed that pastoral management regimes can be formalised into effective management bodies capable of cooperating with government and outsiders in both decision-making and maintenance of assets. This has proved to be wrong in Ethiopia, where donor-financed ponds have not been maintained due to poor siting and lack of ownership. Behnke (1995) argues that this is partly because African pastoral tenure systems are mistakenly assumed to be common pool resource management regimes. He argues that they are instead hybrid systems of collective management which evolved in pre-industrial days, and that common pool theorising is therefore of little relevance to them.

Overall, then, problems of lack of information (and misinterpretation) make reform of pastoral management regimes as problematic as the maintenance of the status quo. However, what is clear is that government policy for the late 1990s and beyond should be based on the goal of sustainable rangeland production, not rangeland conservation, and that some form of joint management, whereby the state underwrites security of tenure for indigenous producers who manage according to their own customs, will be required.

# 3

# REFORMING THE STATE

Within the public sector, two parallel processes of reform have been underway during the late 1980s and early 1990s. The first has involved the scaling down of public sector involvement in natural resource management and agricultural service provision. Ideally this is well-planned and gradually phased withdrawal, although this has not always been the case. The second process of reform has involved changes, often of working practices or the locus of decision-making power, within what remains of the public sector so as to make it more efficient. Key features of the second type of reform have been attempts to increase accountability to rural people and the ability of public servants to respond to rural people's needs in a flexible and effective way. In certain cases this has required a redefinition of the respective roles of government departments (often from a direct involvement in provision to a more regulatory or facilitating role) and a strengthening of the links between key ministries with a stake in the natural resource area.

## Types of change

### Scaling down of public sector efforts

International lenders have placed a great emphasis on reducing the absolute numbers of public sector staff, as a cost-cutting measure. Where privatisation is the underlying philosophy of reform, down-sizing of the public sector should be possible since ownership is directly transferred out of the public sector. For example, the management of numerous irrigation schemes throughout the world has been handed over to groups of users or specially created companies. Similarly the progress towards privatisation of some agricultural marketing systems has been impressive. In Tanzania, for example, components of the National Milling Corporation were privatised to the extent that its residual share of marketed maize dropped from 90 per cent to 2 per cent (Mans, 1994).

Despite these apparent successes, however, cutting the number of civil servants has not proved an easy task. Progress is slow, and gains have sometimes been reversed when those that were laid off were re-employed (World Bank, 1997).

In focusing on this aspect of reform, donors and lenders have tried to alter the underlying philosophy of states, turning them away from providing guaranteed employment for all graduates or well-connected people and towards needs-based employment and promotion based on merit. One way in which this has been done is to hive off sections of former ministries to create bodies similar to the 'quangos' (quasi autonomous non-governmental organisations) which are familiar in the UK. In Costa Rica, for example, the European Union has sponsored the formation of such a hybrid organisation which is able to operate with much more flexible employment practices than full government bodies, hiring staff on contracts as the need arises (Richards *et al.*, 1996). Even when a decline in numbers is achieved, this does not always result in declining costs to the government. Remaining state employees (or state-sponsored quango employees) often need to be better resourced and better paid so that they are more motivated and better able to perform in their jobs. Certainly, Chowdhury and Gilbert (1996) report that massive restructuring in the financially paralysed Bangladesh extension service did not reduce overall costs.

### Efforts to enhance the efficiency of the state

#### Decentralisation of government

To its supporters, decentralisation puts decision making in the hands of people who are well-informed, accessible to others, and in a position to make decisions, which are fundamental to the lives of many rural people, in a timely manner. The most common approach is to deconcentrate central government authority so that regional or even local level authorities can act in a more autonomous manner (though guidelines are still set from the centre). The underlying premise of this type of reform is that physical proximity to users is likely to result in better informed, more effective and more transparent public-sector decision making. This in turn should mean that covert relationships and hidden patronage relations are revealed, to the benefit of the larger population of users and at the cost of the individuals who formerly benefited from them (Hobley, 1995). The success of such reform depends on a number of factors, including whether mechanisms are established for linking with local user groups, whether financial authority is also deconcentrated in order to ensure that decisions can be operationalised, and whether lower level staff have the skills and capacity to manage the new tasks for which they are responsible (Bardhan, 1997; Carney, 1995b).

All these areas appear to have been addressed within the Bangladesh extension reform programme. Although this started rather slowly, it has now begun to be effective. Financial control and a mandate to spend in response to locally-identified needs, rather than in pursuit of predesigned programmes, has been delegated to district and thana levels. Procedures have been put in place to support local-level planning, to assess users' needs and to target a certain portion of expenditure towards women farmers, marginal farmers and landless people. Comprehensive retraining programmes for the 13,000 government extension staff have been put in place. Extension officers have gained not only from such formal training exercises but also from the fact that they are now mandated to respond directly to farmers' needs; their credibility among farmers has thus risen substantially.

### Change in role of the public sector

As the debate over privatisation has progressed, commentary has begun increasingly to focus on the residual government responsibility for regulation. When competition is introduced through privatisation, regulation may not be vital (except in the general sense of ensuring the enforceability of contracts and where significant externalities exist – see below). However, when what are effectively local or national monopolies are passed over into the private sector, effective regulation is absolutely critical if the full benefits of private ownership are to be captured (Kirkpatrick, 1996). By their very nature and spatial dimensions, the management and supply of many natural resources have monopolistic features. This means that continued government monitoring (without needless intervention) is highly desirable, particularly to ensure that users – or certain groups of users – are not disadvantaged.

The need for regulation is further enhanced by certain economic characteristics of various natural resources. Where significant externalities exist government is likely to have to monitor performance and intervene both by setting standards and by ensuring that these are adhered to. Thus, for example, the government must protect producers by maintaining phytosanitary regulations and protect consumers by laying down water quality standards and prohibiting the use of excessively dangerous agrochemicals. It must also look at the climate, erosion and biodiversity implications of felling forests. On the other hand, in some areas (such as the testing, official release and popularisation of new seeds) regulations are often overly restrictive and require reform if wider choice is to be provided to farmers (Tripp and Gisselquist, 1996).

One notable reform in a number of countries in Latin America has been explicitly to recognise the environmental dimensions inherent in natural resource management. In Mexico (1982), Brazil (1992) and Honduras (1993) new ministries of the environment have been established, which to some extent take on the role of the regulator (Richards, 1995). The performance of such ministries varies with context. Richards *et al.* (1996) contrast

23

the approach in Mexico (decentralisation and deregulation) with Costa Rica (decentralisation and state regulation), and find the latter to be a more promising approach. However, in general, examples of effective government regulation are lacking. One reason for this is the fact that the skills required for regulation are even greater than those required for direct provision or management (Batley, 1996). The role of the regulator is much more subtle, and involves far more points of interaction with the private sector whose motivations are – by the nature of regulation – not always closely aligned with perceived public interest.

### Increased coordination within the public sector

In most developing countries, a good deal of waste or mismanagement has resulted from poor definition of the responsibilities of different ministries and a lack of contact between individuals in different line departments. Probably the most widely discussed example of this lies in the lack of linkages between agricultural research and extension services (Kaimowitz, 1990). One of the most common approaches to improving such linkages has been to form joint committees to bring the various stakeholding departments together. For example, in Bangladesh, Agricultural Technical Committees bring together researchers and extensionists in an attempt to ensure that research programmes are developed in response to local needs.

In other cases, the mandates of different departments conflict in ways which may only be revealed during the reform process itself. During the implementation of a programme of irrigation management transfer in Indonesia, it became apparent that the Ministry of Public Works, which oversees irrigation, was not mandated to collect money. This responsibility was vested in the Ministry of Home Affairs. It was therefore the reform process itself that brought these two ministries together to devise a workable strategy, which has resulted in close cooperation at a local level.

More sweeping, cross-sectoral reform has been witnessed in Costa Rica where the Ministry of Natural Resources, Energy and Mines was created following the election of President Arias in 1986. This ministry brought together the National Park Service, the Directorate of Wildlife, the General Directorate of Forestry and six smaller environmental and economic agencies so that forestry policy could be addressed in the context of wider environmental considerations (Richards et al., 1996).[1]

Another approach to the problem has been to monetise the transactions between different departments. The principle here is that if one department depends upon another for resources (through contracts), then the two will work together in a cost-effective way. In some cases, non-public sector bodies can also compete for the contracts. Once again, in Bangladesh, over twenty-eight contracts have been signed between the Department of Agricultural Extension and various public sector research institutes (Walker, 1996).

## Increased regionalisation of effort

Regionalisation of effort so that various countries pool resources (or jointly seek donor support) to address a shared concern can, or should be able to, increase the efficiency of public sector expenditure. This opportunity is most often discussed in the context of agricultural research: costs are high, many problems are common to a number of contiguous countries and economies of scope and scale should be significant. The benefits should be greatest for 'small' countries, which may otherwise fail to reach minimum efficient scale in research efforts. These countries need to work together and to become intelligent 'borrowers' of technology, which also implies that they need to have good information about research efforts elsewhere and efficient screening procedures (Eyzaguirre, 1996).

In October 1996 the Global Forum on Agricultural Research held its inaugural meeting (attached to the International Centres Week of the CGIAR (Consultative Group on International Agricultural Research)). Among other matters relating to the needs and opportunities for agricultural research, the notion of building up strong regional research forums was debated by agricultural scientists, NGOs and farmers' organisation representatives. In fact, such regional organisations have existed for the West Asia and North Africa and the Asian region for more than a decade, though they have had relatively little impact. At the meeting the Latin American and Caribbean and sub-Saharan African regions agreed to establish coordinating mechanisms. Much of the effort will, though, be focused at the sub-regional level where organisations such as CORAF (the Conférence des Responsables de la Recherche Agronomique Africaine) in West Africa and ASARECA (the Association for Strengthening Agricultural Research in Eastern and Southern Africa) have been achieving renewed prominence of late.

Overall, then, reform within the public sector has aimed at streamlining activities, reducing costs and rationalising decision-making so that it becomes more needs-based.

## Problems encountered in reforming the state

Reform, particularly of a radical nature, is never straightforward. The aim of the research upon which this book is based has been to draw out lessons from early experiences of reform. Many of these lessons are derived as much from the problems which have been encountered in programmes of reform as from the successes. The purpose of this section is therefore to highlight some of these problems and, in conclusion, to suggest some elements which are likely to be critical to future reform programmes.

## Resistance to change

Many of the reforms detailed above appear threatening to those working in the public sector. The threat can be real and the fears well-founded; the overall primacy of the state is being reduced and, at the same time, individual public servants are being required to take on new tasks which they may be ill-equipped to fulfil. The explicit goal of reform is to raise the status of clients and non-public sector partners in the management of natural resources and supply of agricultural services. New extension paradigms embrace the concept that farmers should have real decision-making power, new joint forest management schemes explicitly reduce the relative power of the state and formalise the participation of user groups, and irrigation management transfer programmes take many decisions about spending out of the hands of irrigation departments (so removing numerous opportunities for patronage).

Furthermore, decentralisation within the public sector implies that those at the top cede decision-making power, and (if well-conceived) financial control, to those lower down in the system. It is only to be expected that such changes will engender resistance from within the system, especially since it is those at the top of existing structures who are most threatened by reform, yet it is frequently these people who must take responsibility for driving reform through.

Resistance to change can take a variety of forms. It can involve stalling over the implementation of changes by groups or individuals within the system. The Forest Department in the Indian State of Haryana, for example, has refused to enlist other state agencies in implementation of its Joint Forest Management (JFM) programme as it fears losing ownership. It has thus failed to implement a number of changes. Also in Haryana, the Department of Rural Development and Panchayats has refused to sanction the passage of a combined resolution for social forestry and JFM which was forwarded to the Government of Haryana in 1990 to be issued as a government resolution. The Department feared that, should the resolution be issued, its power base would be undermined (Hobley and Shah, 1996b).

Resistance can also mean that change is never really operationalised; while all the superficial changes may have been made, the complex underlying relationships may remain unaltered. District forestry officers in Haryana reject decisions which have been made, not by voicing opposition but by failing to enforce the directives. In particular, they have resisted the transfer of power to communities because this threatens their own position. The same was true in the Costa Rican forest sector where new Regional Committees were created but, because they were not given adequate resources and effective decision-making power, they became moribund. The same has been true to some extent of the various regional agricultural research initiatives which were established in the 1980s. National research systems have been reluctant to cede to them resources and decision-making power. As a result they have been emasculated, although the new Global Forum on Agricultural Research is attempting to reverse this.

Other efforts have been made to overcome the problems. For example, in Rajasthan new funds have been made available for which public sector agricultural researchers are expected to compete. This is supposed to provide a concrete mechanism to enforce a change in working practices. However, since government researchers are still evaluated on their scientific performance rather than their ability to assist farmers in meeting their needs, there is inadequate incentive for them to engage in the competitive contracting process. Changes in responsiveness to clients' needs generated through new sources of finance are therefore unlikely to be fully effective unless supported by other types of internal reform, especially in staff performance assessment criteria and reward systems.

In other cases, however, it seems to be lack of changes in financing arrangements which have been the root cause of the problem. Central authorities are often reluctant to devolve financial control (and the associated powers of patronage) to lower levels. Part of the problem is that the kind of changes which are being promoted require individuals to take risks, to learn from experience and to admit failures. Bureaucratic structures, on the other hand, tend to reward risk-averse people who conform to the norms of their institutions. It is also a tendency of managers to refuse to let their subordinates (and new partners outside government) learn from their own mistakes. This can lead to excessive intervention in decision-making at lower levels (Hobley and Shah, 1996a).

Reform can also be hijacked by political forces within a system. Ultimately bureaucracies are (or should be) responsible to the elected representatives of the people. It is frequently the case that these politicians (and the power they wield) are neglected during the early stages of policy formulation, especially in reform programmes which are donor sponsored. It is usually not sufficient simply to seek agreement on the purpose of a reform. More time needs to be invested in detailing the specific activities and the implications of any vision for institutional change (at the outset).

For example, in Kenya the donors pushed forward a process of reform of the parastatal agricultural marketing system. While they engaged with the bureaucracy and the parastatals themselves, they failed to court political support for the reform, with the result that the reform programme achieved only limited success (Lewa, 1995). Similarly, in Bangladesh, the reform of the extension system has not yet been tied to reform of the political system in the country, including the devolution of power to locally elected political bodies. This leaves national level politicians with powers to override changes if they so wish (Chowdhury and Gilbert, 1996). The same problem has been observed in the water sub-sector, where Turral (1995b) concludes that few countries have paid systematic attention to all stages of the reform process. Wide-ranging political support has not generally been achieved, and in many cases the creation of a favourable environment for transfer of irrigation management out of the public sector has not preceded implementation.

This underlines the fact that political support may have to be sought beyond the immediate sub-sector in which reform is taking place. Our research repeatedly demonstrates that reforms are far more likely to be successful if they take place in a favourable economic environment in which markets are well-developed. For example, groups in Haryana successfully managed *bhabbar* grass (*Eulialopsis binata*) leases because the grass was a high value resource which was relatively easy to market. Similarly, community-based forest management has proved easier to sustain in more valuable areas (areas with a higher density of mahogany and other such trees; Richards, 1995).

The creation of such an enabling economic environment is a complex task, and clearly beyond the scope of the Ministry of Agriculture or Forestry alone. It is, however, an area to which donors in general are paying more attention. High levels of failure or sub-standard performance of past development projects and aid spending more broadly have focused attention on the need to work only with countries which have already exhibited concrete commitment to public sector reform and creating an enabling environment for pluralistic development. Thus, it has been argued that agriculture sector aid should flow only to those countries which demonstrate a clear commitment to agricultural development by opening up national debate on priorities, incorporating farmers into decision-making an allocating sufficient domestic resources to the sector (World Bank Africa Region, 1995).

Finally, reform within government institutions can be opposed by individuals outside the government who benefited from pre-reform structures and configurations of interests, whether legitimate or not. Hobley (1995) cites the example of forest department officials who receive direct payments from influential villagers for privileged access to forest resources. Reform designed to expose such practices is likely to be resisted by politically powerful groups which have paid large sums of money over time to establish their position. Where such individuals also dominate elected and appointed local level decision-making forums, change will be difficult to implement (Mosse, 1996).

## Lack of shared vision within and between ministries

The existence of this problem is one reason why there is such great need to court wide-ranging political support for reform. Decentralisation of responsibility, by its very nature, means that change needs to take place at a large number of different points within the system. In India, water is a state (as opposed to federal) responsibility, which means that there are numerous points at which reforms officially approved at national level can be derailed. For example, national water law and national legislation on groundwater has yet to be ratified by more than a handful of states. Furthermore, at state level numerous different bodies can be involved in policy formulation and implementation; in Gujarat, there are at least four statutory bodies within the public sector with responsibilities for water issues.

Bold new ideas about bringing together different ministries or departments with synergistic areas of operation have proved more difficult to implement than might have been thought. For example, the various departments within Costa Rica's new Ministry of Natural Resources, Energy and Mines were found to suffer from conflicting aims and overlapping remits. Unresolved legal issues further exacerbated problems and prevented coordination and the efficient use of resources. Also in Costa Rica, within the Tortugero Conservation area, lack of cooperation between different ministries has persisted, despite the signing of a series of inter-ministerial *convenios* (agreements). The Institute of Agrarian Development has continued to hand out large forested areas to colonising small farmers and the Ministry of Agriculture has supported the expansion of banana plantations in direct conflict with (decentralised) project objectives. Paradoxically, these banana companies are themselves sponsored at national level by the same ministry which objects to their plans at regional level (Richards *et al.*, 1996).

These examples seem to give weight to Grindle and Hilderbrand's (1995) contention that where serious problems exist in communication and coordination between organisations, 'tinkering with organisation charts' does not produce effective results. Rather, active coordination mechanisms, which they argue should be 'task specific', need to be sought. This also illustrates the importance of decentralising resources so that local-level decision-making bodies can implement their own decisions and challenge the centre if they see fit. In this context, it is interesting to note that in New Zealand the functions of the Forest Ministry have been separated on the grounds that inter-ministerial wrangling is more desirable, transparent, and ultimately more accountable than intra-ministerial wrangling (Hobley, 1995).[2] This will, however, only be true if fragmented departments have sufficient interest at stake that they chose 'voice' (or conflict) over 'exit' (or ceding responsibility).

Efforts to sponsor linkages between agricultural research and extension are still underway in many countries. Certainly no universally applicable model for achieving this task has thus far been promulgated, and the two public sector sides of the 'technology triangle' (farmers represent the third 'side') remain distant in many countries (Merrill-Sands and Kaimowitz, 1991). In Bangladesh, reform has been hampered by a number of factors including a lack of capacity (in terms of numbers of researchers in each area) within the research system and a reluctance by researchers to attribute equal priority to needs articulated through extension officers and those identified by their own colleagues. Early ideas that formal farmers' organisations might be able to use their ability to apply pressure and their political influence to ensure that research and extension departments work together in the interests of small farmers have yielded little evidence in their support (Carney, 1996a). Such farmers' organisations lack technical and financial capacity and rarely seem to view technology as a priority area of engagement. This is not to say that they might not move more deeply into the technology area at a later date, but this is not usually a priority for 'young' organisations. These must focus on consolidating their own existence before tackling an area as potentially costly, risky and divisive as technology development.

This appears to place responsibility for the effectiveness of the agricultural technology system firmly back in the hands of the public sector, and thus seems to call for further refinements in thinking about reform. Drawing on field research in six developing countries, Grindle and Hilderbrand (1995: 441), for example, argue that public sector performance is 'more often driven by strong organisational cultures, good management practices, and effective communication networks than it is by rules, regulations or procedures and pay scales'. They also note that capacity-building projects have so far generated few benefits, and that alternative solutions must be found. One option that is gaining credence is to move away from long-term expatriate technical assistance for capacity-building and to move towards supplying resources for shorter term consultancies on demand (using either domestic or international consultants as appropriate) (OECD/DAC, 1997).

### Incomplete or ineffectual reform

There are numerous instances of incomplete or ineffectual reform in all sectors. Problems may occur simply because all the dimensions of reform are not adequately considered at the outset. For example, reform in the Costa Rican forest sector did not take into account necessary changes in the legal structure (Richards *et al.*, 1996). A similar problem is echoed by Cox and Behnke (1995), who note that contradictions and confusion in legal provisions for land enclosure are the major impediment to improvements in rangeland management in Namibia. Analogous stumbling blocks occur when reform inadvertently increases (rather then decreases) bureaucratic procedures. Again in Costa Rica, overly bureaucratic requirements for seeking cutting permits from the government are reputed to have encouraged illegal felling of trees (Richards *et al.*, 1996).

Hobley (1995) makes a somewhat more damning argument. She holds that recent changes in the forest sector in India have tended to increase the power of the state at village level, rather than to decentralise power to local people. It is her view that the power of covert institutions remains dominant and that relations within formal institutions are still conditioned by patronage and rent-seeking behaviour. There are a number of reasons for this, many of which relate to unequal relations at local level and the fact that inadequate efforts have been put into addressing these. In particular, little effort has been put into improving communications between field-level officers and their immediate superiors, or into monitoring the process of reform and using the information gained to improve the design of the reform programmes (Hobley and Shah, 1996a).

## Capacity problems

The success of much reform within state organisations hinges to a great extent on the capacity of the public sector to embrace change and to alter its working practices. Frequently, prior to reform, it was a lack of skills within the public sector which resulted in such poor levels of service or management. The difficulty faced by proponents of reform is that the new roles which they have defined for public servants require even greater skill levels than the previous systems. In T&V systems, extension officers effectively had the task of communicating predetermined messages to contact farmers. Now they must make their own decisions about appropriate messages and act as a two-way communication channel between farmers and researchers. This has not been easy. Monitoring of extension officers in Bangladesh during 1993–4 revealed many problems with the demonstrations and trials which extension officers had been given the task of establishing in response to farmers' needs. Since they had inadequate training and experience, the extension officers were unable to select appropriate sites and design and implement the trials in an effective manner. This is a critical issue, since once local extension workers lose credibility with farmers it can be very difficult to recover it. An additional problem in Bangladesh concerned the monitoring itself, and the capacity of the state to assess which technologies had been successfully adopted at a local level and subsequently to transfer lessons to other areas of the country. An inability to do this significantly reduces the effectiveness of any extension service.

In general, supporters of decentralisation applaud the expansion in the number of decision-makers within the public sector which it entails. However, critics claim – with some justification – that this necessarily lowers the overall quality of decision-making within government. They argue that only at the centre are there individuals of sufficient capacity and experience to understand the full implications of their decision-making and consequently to learn from the mistakes and success of others. Evidence for such a view can be gathered from a country such as China, where the small proportion of qualified people within the extension system are concentrated at the top of the administrative structure. At provincial level, 77 per cent of employees have a college education. At county level the proportion falls to 36 per cent, then to 17 per cent at town level. This translates into an average of one person with college-level education for every six towns, and less than one person per town with school-level agricultural education. At village level there is essentially nobody with any formal agricultural education (CECAT/RCRE, 1996). This problem is likely to be exacerbated in the early phases of public sector reform, as those who are well-qualified often choose to leave government employment and respond to more lucrative private sector opportunities which open up as the public sector contracts.

It is certainly not the case that lower-level authorities always have the best interests of local people at heart. They too can be dominated by elites and are prone to use their powers for patronage rather than broad-based, equitable change (although the opportunities for patronage and rent-seeking tend also to be more limited in decentralised systems: there are few multi-million dollar infrastructure investments, for example). Some would, however, argue that local authority personnel are more prone to such forces than those from central authorities since they are more accessible to powerful pressure groups or individuals. This links back to the issue of the importance of courting a political constituency for change at all levels, and ensuring that knowledge of the details of reform programmes is widely dispersed to limit abuse. Crook and Manor (1994) report that perceptions of corruption rose in the newly decentralised system of government in Karnataka, India, although this was due more to increased openness than an actual rise in errant behaviour.

One of the most prominent role changes in most reform programmes is that from public servants managing resources and the supply of services themselves to acting as regulators and possibly financers of services provided by others, often under competitive bidding processes. When irrigation management transfer takes place, public agencies must change from being system managers to overseeing functions such as the registration of water rights, the collection, analysis and provision of hydrological and water quality data, regulation, and enforcement of environmental codes. These are notoriously difficult tasks and are likely to require substantial retraining of irrigation department personnel if they are to be effectively performed (Turral, 1995b). Indeed, there are very few examples of effective regulation driven by the public sector in developing countries. This problem is one aspect of the 'orthodox paradox', which is the term which has been given to the dilemma posed by calls for a minimalist role for the state, coupled with an effective state apparatus to see reform through and regulate the new environment (Grindle and Hilderbrand, 1995).

The upgrading of skills is also important if local-level agents are to be in a position to elicit and devise programmes which are responsive to local needs. Demand articulation is not as automatic as might be assumed, and NGOs in particular have promoted their belief that significant local-level (that is, outside government) capacity building needs to take place, usually through the formation of local resource 'user groups', before needs can be expressed and subsequently prioritised and addressed. If decentralisation within government is designed in large part to institutionalise responsiveness to local people's needs, this is a key issue. It is also one that is frequently ignored when plans are drawn up to institutionalise 'demand driven' development processes. Many extension workers and forest department officials, for example, do not have the social skills required to promote group development. This was certainly true of extension officers in Bangladesh. Furthermore, researchers and extension officers may view group formation as 'below them'. This was true of researchers from the Department of Research and Specialist Service in Zimbabwe, who are working on soil conservation techniques with groups in a DFID-sponsored project (Steve Twomlow, personal communication). Overcoming such prejudices can be very time consuming, but is usually essential if progress is to be made.

A further difficulty, noted in recent work in Rajasthan in relation to the use of research funds, is that NGOs and the farmer groups with which they work rarely have the technical skills to specify agricultural problems in sufficient detail for researchers to then address them (Alsop et al., 1996). To counter this, part of the agricultural research fund in Udaipur District was set aside for NGOs to commission independent advice in the first instance, and to help them to formulate their research requirements.

Finally, it should be noted that many of the new technologies, such as Geographical Information Systems (GIS) and computer-assisted planning, which are rightly becoming available to developing-country decision makers, have within them inherent tendencies towards centralisation, as does the move towards regionalisation of agricultural research effort. It will remain important to ensure that mechanisms are put in place to enable users (farmers' groups, for example) to access more centralised decision-making forums.

### Financing issues

One of the prime motivations behind the reforms of the past decade has been a desire to reduce costs. While theoretical, welfare-economics arguments are often invoked to justify the changes, it is frequently the lack of resources within existing systems which provides the actual impetus for the change process to begin (particularly since donor funds increasingly come with conditions attached). Perhaps one reason why reforms, and particularly state withdrawals, have been less well-planned and coherent than might have been expected is that economic concepts actually provide relatively poor guidance as to the direction or intensity of government withdrawal. Most goods and services combine some aspects of public good and private good characteristics and market imperfections are rife.[3] Indeed, this was the reason why extensive state involvement in developing country economies was originally supported (Colclough, 1991).

Even if economic concepts were able to provide a good guide for reform, problems of rationing public sector expenditure and inter-sectoral priority setting would still remain (Carney, 1995a). Indeed it should always be borne in mind that reform in the natural resources sector runs parallel to reform in all the other sectors in which government has a major commitment (such as health and education), and since all sectors are ultimately interconnected in their impact on individuals' livelihoods, reform overall is correspondingly more difficult. Certainly lessons learned in one sector should be applied to others; too often, sectoral specialists work in isolation.

If the emphasis of reform is on cost reduction, one of the most significant problems faced so far is that there is relatively little evidence that costs have actually fallen. Indeed, in some cases reform has added to public costs (although it might still be considered to have achieved better value for money in that those resources which are spent, are spent better). In Bangladesh, for example, changes within the extension system have increased funding requirements. Other countries have found it extremely difficult to cut employment rosters. Beynon *et al.* (1995) point out that the rate of growth in the numbers of agricultural researchers in sixteen sub-Saharan African countries slowed from 8.3 per cent over the decade of the 1960s to 3.3 per cent in the 1980s, but that in absolute terms, the annual addition to the total number of researchers increased slightly during the 1980s. Civil service reform has in general made relatively little headway against the tides of underpaid but overly numerous bureaucrats in many developing countries.

While donor support has been critical to the process of reform within the public sector, it comes with its own problems. First, it can encourage the creation of systems which are not ultimately financially viable, particularly when donors chose to operate through structures which lie outside existing, often reviled, line ministries. Second, it can introduce damaging rigidities into programmes. For example, coming from a standpoint within western liberal democracies, donors may stipulate that certain target groups must be prioritised (the poor, women, etc.) or that strict ecological/conservationist policies be adopted which may detract from the target of devising a system that is truly responsive to user needs on an ongoing basis (Hobley and Shah, 1996a). Third, it can allow developing countries themselves to avoid painful processes of reallocation of funds from less viable or effective areas of operation to new priority areas. Fourth, and at its most extreme, it can lead to 'aid dependency'. It has been shown that at certain volumes and under certain conditions aid may actively hamper a country's prospects of achieving self-sustaining development (SIDA, 1996).

## Conclusion

It would clearly be unreasonable to expect that reform within the state should have proceeded without any setbacks. Indeed, the evidence gathered here demonstrates how complex reform is, even when only the intra-governmental aspects of the process are considered. When other players are taken into consideration (as they are in the following chapter), relationships become ever more diverse and reform itself becomes even more difficult to manage.

In view of this, it seems to be essential that governments continue criti-cally to analyse their own role, and also to recognise the potential conflicts that they must face. Such conflicts occur between individuals, between departments, between interests within a sub-sector (for example, support of tourism, conservation, rural development and commercial revenue genera-tion are all valid goals for a forest department, but it is unlikely that all can be pursued at once), between sub-sectors (competition for land or water between sedentary farmers, pastoralists and forest dwellers) and between sectors (this is particularly important in China, where agriculture is losing land to industry and housing at a rapid rate). Clear policy pronouncements from government, combined with bold and enforceable statements about the limits to government involvement, are certainly a good starting point if these conflicts are not to be fatal for the reform process. Such a position is supported by Grindle and Hilderbrand (1995), who prioritise 'a sense of mission' and 'commitment to organisational goals among staff' as being of critical importance to public sector performance. These contribute to team-work and the inculcation of a problem-solving attitude as well as to greater coordination between policy and implementation.

It is important to emphasise here that most public sector reform programmes are still ongoing or are not yet underway. While the language of reform has permeated widely among those that are concerned with gover-nance in the natural resources area of developing countries, many hard decisions about the process of reform have yet to be made.

Perhaps, though, greater progress will be made now that the tide of opinion has once more swung somewhat in favour of the state. A heightened awareness of the importance of the state in the development process is reflected in the World Bank's 1997 *World Development Report*. It can only be hoped that reform will gain pace in an environment in which the *merits* of an *effective* public sector are once more stressed. Possibly this will raise morale within the public sector itself and eliminate some of the obstruction of reform caused by fear of total public sector emasculation. Whether or not this is the case, it remains important to monitor reform itself, the processes through which it is introduced and the resulting institutional configurations to ensure that all possible advances towards (social) efficiency, effectiveness and accountability are made.

*Table 1* Reforming the state: a summary

| Principal types of change | Examples and outcomes | Future challenges for each type of change | Generic problems |
|---|---|---|---|
| 1 Scaling down the state. | Irrigation management 'turnover':<br>• Some success, especially in small schemes.<br>• Continuing problems of weaker user groups and 'sabotage' by users uncommitted to groups. | • Develop stronger user groups and improved interfaces between different types of customer.<br>• Ensure local-level staff are resourced and skilled to manage new responsibilities. | • Fear of erosion of authority by state employees and of erosion of power base by influential outsiders often delays or undermines processes of reform. |
| 2 Making the state more efficient: | | | |
| (i) Decentralisation of authority | Agricultural extension in Bangladesh:<br>• Increased responsiveness to local needs.<br>• Local staff command greater respect among farmers. | • Need to enhance central monitoring capacity to learn and spread lessons about the uptake of technologies and performance of different decentralised approaches. | • Performance assessment and reward systems in public sector continue to reward conventional behaviour and potentially penalise risk taking. |
| (ii) Strengthen capacity of state for sensitive regulation of new/ expanded private economic activity. | Ministries of environment created in Mexico, Honduras and Brazil to address environmental issues in forest management. | • Continuing problems of coordination among departments and disciplines. | • Levels of management skill within state organisations are generally inadequate to permit new culture of responding to clients' needs. |
| (iii) Increase coordination within public sector. | Some success in strengthening research-extension linkages in Bangladesh through research contracts and improved liaison committees. | • Performance assessment and reward structures for researchers need to be geared more to client orientation and less to academic criteria.<br>• Importance of generating shared sense of mission. | • Those pursuing administrative reforms rarely seek prior political support.<br>• Donors tend to introduce changes requiring levels of resources unsustainable after their withdrawal. |
| (iv) Increase regionalisation of effort. | New initiatives within the Global Forum on Agricultural Research. | • Ensure that national systems cede adequate authority/resources to supra-national bodies.<br>• Ensure users' needs are represented in international decision-making. | • Costs of scaling up from small-scale efforts are often misjudged. |

# 4

# NON-STATE APPROACHES

The multiplicity of alternative service providers and institutional configurations which have appeared in the agricultural sectors of countries where public sector retrenchment is reasonably advanced is both exciting and daunting. Old style parastatals and government ministries have now been joined, superseded or substituted for by organisations ranging from small-scale farmer groups taking marketing into their own hands, to local subsidiaries of multinational giants supplying hybrid seed to farmers. In between these two extremes lies a continuum of different institutions, some organised in pursuit of financial profit, others in pursuit of social goals, some almost too close to the government organisations they replace, others radically opposed. University departments and unions, small-scale traders and NGOs are all feeling their way through the new environment, probing the boundaries and assessing the chances of fulfilling their at times conflicting objectives (Carney, 1995c).

Like all profound change, this is not a simple process and there are no clear rules as to how it should proceed. In addition, there will be much country-specific variation. However, while there may be no blueprints for the different roles of the various actors, there is a growing pool of evidence and theoretical analysis upon which decision-makers can draw when thinking about institutional configurations for effective resource management and service delivery. This will yield general guidelines and lay out some broad boundaries of the potential roles of different players rather than leading to inviolable prescriptions.

This chapter reviews the different activities undertaken by the four main types of alternative service providers/resource managers within the natural resources sector: common pool resource management groups, membership organisations involved in service provision, non-membership NGOs (which frequently create and work through local-level membership groups), and the private commercial sector. In each case, we examine early evidence of success or failure and draw lessons from this evidence about preconditions for future success.[1]

## Common pool resource management groups[2]

Common pool resource management groups (henceforth CPR groups) have received increasing amounts of attention over the past two to three decades. Early analysis of their activities tended to be pessimistic and unsubtle, in that it was permeated with confusion about the vital differences between open access and common pool resources. The two are distinguished by the regulations which govern their use. The power to exclude outsiders is the essence of common property regimes (local management structures for common pool resources). This means that common pool resources are, in effect, 'private property for the group', which makes their degradation and overuse (which is characteristic of open-access resources) far from inevitable (Bromley, 1989; Behnke, 1995). Debate has now progressed, and one important focus for research is the internal and external conditions which make CPR groups robust.

### *What has been done?*

CPR groups are active in all the areas of natural resource management with which this book is concerned (forest, water and pastoral resource management). Reform in all these sub-sectors, and a renewed emphasis on collective action and user participation more generally has highlighted the importance of CPR groups. It is therefore vital that their strengths, weaknesses and the likely boundaries of their activity should be well understood.

The resources under consideration in this book share three essential features which argue for joint action in resource management:

- although in many cases the *ownership* of forests, rangelands and water resources is vested in the state, the state does not have the capacity to police rules governing their access and use;
- the management of such resources by individual users is problematic, principally because of the difficulty of defining boundaries (in the case of water and rangelands) and the high cost to individuals of policing them (in the case of rangelands and forest);
- private (commercial or individual) management of such resources tends to benefit only one group (for example, the commercial beneficiaries of long-term forest concessions); since most natural resources have multiple users and uses, private management is therefore likely to have unacceptably high negative equity impacts.

These issues are now widely accepted, and management by CPR groups – many of which had been destroyed in their original form by privatisation or nationalisation of resources in the past – is once more encouraged.

Perhaps the most prominent examples of CPR group involvement in resource management come from the forest sector. In India, Joint Forest Management (JFM) has been adopted as an all-India policy (although its implementation is highly uneven) and in Mexico, *ejidos* (cooperative land-holdings created following the Mexican revolution) have been managing local resources since 1918. Reforms in irrigation management have also given new prominence to water user associations, which have been created to take responsibility for small and medium scale irrigation schemes which were previously managed by the state. These now join remaining networks of traditional water users (many of which have developed complex systems of management over time, although some have been eroded with state intervention) as legitimate and 'acceptable' CPR groups (Pant and Cox, 1996).

The situation with regard to rangeland management is, according to Behnke (1995), somewhat more complex. Where groups of users are externally regulated by government, they can be considered to be CPR groups and analysis of their activities can therefore enrich our overall understanding of the dynamics of such groups. However, where there is no state regulation or intervention, Behnke concludes that groups of pastoralists cannot be considered to be CPR groups in the classical sense. Pastoral groups adhere to constantly shifting usage rights and regulations which have proved to be both resilient and successful in managing access but which, 'with remarkable regularity', do not fulfil the conditions of successful common property regimes outlined below. Key reasons for this are the fact that they entail mixed, rather than universally communal, tenure regimes and that resource boundaries are permeable and ill-defined.

### Outcomes and why

There seems to be broad agreement that, when groups function as intended, common property regimes have been more effective than the state in managing resources. Unlike government bodies, CPR groups have a direct stake in the future of the resource which they are managing, they have ready access to information about members' needs, and they can react quickly to changes. This means that both their incentives for good management and their ability to manage well are enhanced (see the section on preconditions for success, below, for conditions which must apply).

However, not all the outcomes of attempts to encourage CPR management have been positive. Sometimes this is because genuine efforts to hand over decisions about management are not made (the problems of transfer of responsibility outside government are as great, if not greater, than the problems of decentralisation within government which were reviewed in Chapter 3). For example, many of the village forest committees which have been established under India's JFM programme are in reality little more than an arm of the Forest Department. Since they depend for their existence on the Forest Department, they are certainly in no position to challenge it. Even where groups do have independent standing, there are a number of common problems or challenges which need to be addressed.

### Defining group membership

The philosophy behind CPR groups is that users and managers of resources should be the same people (although users may receive external support for their management activities from non-users). This raises two immediate problems. The first relates to the definition of 'users', and the second to the dynamics of management. Most resources are put to a multitude of different uses and 'normal' indicators of use (such as frequency of use, proximity of resource to users) may not be helpful in defining use rights. Infrequent users (such as medicinal herb collectors) may depend on the resource for their livelihoods, while frequent users (such as women who gather firewood for home use) may be relatively less dependent on it in terms of their overall livelihoods. Likewise, some traditional users of a resource may live relatively far from it. Problems also arise when, for example, the owners of land are not the users of that land and the resources associated with it. Turral (1995b) cites an example from Rajasthan where irrigators were not land owners and they were therefore legally prevented from being members of the local water user associations. Finally, it should be noted that the externalities associated with particular resource management practices can be high (for example, the erosion effects relating to felling of trees), implying that attention should also be paid to the needs of 'secondary users' who might not have a direct stake in management.

Joint forest management and other common resource management policies which entail outside intervention require decisions to be made on these complex matters. A first step might be to bring all stakeholders together to discuss the issues among themselves. This may resolve some problems, but then begs the question of who the stakeholders are (Ravnborg and Ashby, 1996). Turral (1995b) notes relatively little success in tackling this problem. He believes that groups formed on large-scale irrigation schemes are still far from being fully representative of the body of water users. In many cases, groups were set up by irrigation authorities in a perfunctory fashion and have not survived; in other cases, they do not represent more than 10 per cent of users, creating major scope for individual action which is incompatible with group aims.

Once users have been defined, there remains the problem of how relations between the various factions and sub-groups are to be regulated. As Sarin (1996) notes: 'communities are not homogenous but are differentiated by caste, class, tribe, religion, ethnicity and gender with each group often having a specific pattern of interaction with the local resource endowment'. CPR groups which are formed without outside intervention tend to replicate and legitimise the existing hierarchy between sub-groups within a commu-

40

nity. When outsiders intervene in group formation, they can alter the 'natural' pattern of the hierarchy, though there is still a tendency for some of its dimensions to be replicated, since sub-groups which are 'voiceless' tend to be neglected by planners who are unfamiliar with local social relations (Mosse, 1994 and 1996). A second-order question which external agents must then address is whether attempts to challenge existing configurations of interests will result in unsustainable – or overly project dependent – groups. This appears to have been the case in various past interventions in villages in tribal west India (Mosse, 1996).

### Equity and participation

It is common to assume that local people will always be willing to participate in the management of a resource, despite considerable evidence to the contrary. Willingness to participate depends upon a number of factors, including the costs and potential benefits of doing so, and the overall sense of responsibility or ownership felt by a CPR group and the individuals who comprise it. Kerven and Cox (1996) point out that participation must be invited early in the planning process if ownership (of pools of drinking water for the use of pastoralists in Ethiopia, in their case) is to be instilled in group members. Certainly, simply nominating somebody to a group or committee does not constitute participation or ensure that person's views will be heeded.

In many Indian JFM programmes, it has been demonstrated that forest users are unaware of the rights that are conferred upon them by membership of the JFM committee. Alternatively, such rights may be outweighed by complex personal constraints and relationships which, somewhat paradoxically, may be more intense and more limiting the more intimate the group is (for example, family members may constrain a representative's behaviour within a particular group for reasons of social convention) (Mosse, 1996). Another tendency is for people simply to elect leaders who will take decisions on their behalf without a prior process of consultation. Changing such attitudes can take years and requires, among other things, thought to be given to procedural aspects such as the timing and venue of meetings so that poorer groups are not disadvantaged (Hobley, 1995). Indeed, the relative inability of poor people to spend long periods of time in participatory planning sessions is one issue which has been too often overlooked by advocates of intensive participatory methodologies.

Capacity for management is also frequently overlooked. Turral (1995b) notes the complexity of managing irrigation systems and asserts that the skills of group members need to be developed. They are likely to need training not only in negotiation and conflict resolution, but also in areas such as record keeping and the maintenance of simple accounts. NGOs might be able to provide such training, though they too may have few skills in such new areas. It is therefore important that consideration is given to training and who might be able to provide it when reforms are planned.

## Legal framework and enabling environment

It is not only – indeed, possibly not primarily – group level characteristics which affect management capability. Poorly specified or unenforceable legal regulations can have a profound effect on groups. In India, this problem is once again compounded by the fact that responsibility for many legal issues is shared between the federal government and individual states.

Legal reform is not, however, a simple matter. Behnke (1995) argues that specifying rights in any format (in other words, formalising existing rights or superimposing new legal ownership frameworks) is problematic and unlikely to solve the problems of management in that it may well increase, rather than reduce, uncertainty at least for a considerable period. Similarly, Richards *et al.* (1996) note that reforms in Mexico to allow local groups (*ejidos*) alienable land ownership, rather than just ownership of the trees, could encourage groups to clear and sell their land as individual plots rather than to manage it as a common resource. This has already happened with agricultural land. Hobley and Shah (1996b) conclude that groups must participate not only in aspects of management but also in defining the rules which govern their activity. However, the authors point out that deciding how much legal authority the state has over user groups remains a delicate issue. In Haryana, the state has set detailed provisions for the constitution and functioning of user groups. These are intolerant of idiosyncrasy and so are more regulating in character than facilitating. This can in turn reduce members' sense of 'ownership', and thus the vibrancy of the groups themselves.

Another critical factor is access to financial resources. Turral (1995b) notes that water user groups have been particularly weak when called upon to raise money to pay for repairs and maintenance (and that they have implicitly assumed – with some justification – that the government will ultimately provide the necessary support).

It is logical that users will be more likely to be reluctant to make financial contributions when they anticipate state support at the eleventh hour. However, reluctance to contribute may be a sign of actual inability to contribute, given lack of access to savings or credit facilities and high levels of rural poverty. Much agricultural credit was formerly provided by parastatal enterprises responsible for integrated marketing and input supply services. Poor repayment rates ultimately meant that a good deal of such credit was effectively grant financing to farmers.

Clearly, such a situation was unsustainable and, with the opening up of marketing services, the supply of 'easy' credit has entirely dried up in many rural areas. Non-governmental organisations and, in a few cases, rural banks have expanded their operations in a number of places, but there are still large areas of the developing world in which no credit is available from the (semi-) formal sector. At the same time there are no viable saving facilities, so farmers find it hard to plan for lump sum contributions. This problem is widespread, but is clearly more acute in the most marginal areas where ability to repay credit at interest rates which come close to covering costs (which tend to be higher in such areas because of spatial dispersion, low levels of literacy, etc.) is very limited. In principle, those with access to irrigation schemes should be in a better position to take and repay credit in a timely manner and thus to contribute to maintenance costs. However, would-be non-public sector credit providers can be deterred from providing credit even in relatively favoured areas if there is a long history of poor credit repayment.

Related to access to credit is access to markets. Shikavoti (1994) cites examples of insufficient demand for and the difficulty of marketing high-value produce as a limitation on the development of autonomous water management systems in Nepal. In contrast, Hobley and Shah (1996a) found that access to *bhabbar* grass (*Eulialopsis binala*) leases has been the critical cement for CPR groups in Haryana. On the other hand, Richards (1997b) notes that some indigenous groups find the fickleness of market forces very hard to deal with. In Latin America, it is those groups which have built up an acquaintance with markets over a long period (for example, among the Maya of southeast Mexico, those involved in selling *chicle* – derived from *Manilkara zapota* – for chewing gum) which have been the most successful. This militates towards state or NGO support for market entry (in a manner which is in tune with indigenous customs), an issue which has, as yet, received insufficient attention.

### Preconditions for success

The long-term commitment needed for CPR management means that stability and robustness are essential qualities for CPR management groups. Important contributions to the debate that has emerged around these issues have been made by Wade (1988), Ostrom (1990) and McKean (1995). Hobley and Shah (1996b) build on these to compile a set of principles, the fulfilment of which helps to determine the robustness of groups. These are detailed in Box 3.

In terms of government support for CPR groups, there is still probably insufficient empirical material available on the way in which existing groups function for full proposals to be outlined. In particular, future research must address the way in which common pool resources fit into users' overall livelihood strategies. It must also look more closely at notions of 'success', as setting criteria for success against which the performance of such groups can be assessed is a notoriously difficult task and one which commentators have shied away from. Turral (1995b) takes examples from irrigation management transfer and notes that success can be defined in terms of increased cropping area or intensity, rising real per capita incomes, improved water use efficiency, improved productivity per unit of water, or improved equity and access to irrigation. Potentially, all these indicators might lead in different directions. Yet, without agreed determinants of success, it is hard to judge not only the performance of groups, but also the progress achieved by the reforms in their entirety. This latter is especially difficult to interpret because dogma, not firm evidence, has governed many decisions for change.

Furthermore, successes achieved on a small scale may not be replicated as the scale of operations increases. Certainly, many of the early 'lessons' relating to common management of water resources were derived from countries or areas in which there was relatively little land under irrigation, for example, Sri Lanka (0.5m ha), Nepal (0.35m ha) and the Philippines (1.5m ha). By contrast, Uttar Pradesh in India has 14.4m ha. under irrigation. Similarly, lessons derived from experience with small farmer-managed systems have increasingly been applied to large-scale public irrigation systems without adequate thought being given to the vital differences between the two. Overall, the question of scale and how to facilitate and oversee local management or participation over large areas remains largely unaddressed, as does the issue of how best to scale up from pilot initiatives. It has become increasingly apparent that unless such pilots are designed with a view to subsequent replication on a large scale, they may well fail to serve their purpose. A particular concern is that pilot efforts tend to be exceptionally resource intensive which 'prices them out' of contention for wide-scale replication.

---

*Box 3* Criteria for assessing robustness of local forest management institutions

USER GROUP

- *Size*: the smaller the number of users, the better the chances of success. Group agreement is more likely to collapse where there are more than 30–40 members, but tasks cannot be performed effectively once group size falls below a minimum threshold.
- *Boundaries*: the more clearly the boundaries of the group are defined, the better the chances of success. Individuals or households with rights to withdraw resource units from the common pool resource are clearly defined and agreed.
- *Relative power of sub-groups*: the greater the power of those that benefit from retaining the commons, and the greater the weakness

*continued* . . .

---

of those that favour sub-group enclosure or private property, the better the chances of success.

- *Existing arrangements for discussion of common problems*: the better developed these are, the greater the chances of success.
- *Extent to which users are bound by mutual obligations*: the greater the importance of social reputation, the better the chances of success.
- *Punishments against rule-breaking*: the more the users already have enforceable rules elsewhere, the better the chances of success.
- *Consensus about who are the users*: in terms of both customary and legal user rights, this is essential prior to collective action.
- *Distribution of decision-making rights* and use rights to co-owners of the resource need not be egalitarian, but must be considered fair.

## RELATIONSHIP BETWEEN RESOURCES AND USER GROUP

- *Location*: the greater the overlap between the location of the common pool resources and the residence of users, the greater the chance of success.
- *Users' demands*: the greater the demands (up to a limit) and the more vital the resource for survival, the greater the chances of success.
- *Resource capacity*: the better the prospects of multiple benefits in the short term, the greater the chances of success.
- *Users' knowledge*: the better their knowledge of the level of yield that is sustainable, the greater the chance of success.

## INVESTMENT IN THE RESOURCE

Investment in the resource leads to a greater incentive to protect only as long as there is confidence in retaining control over future benefits.

## CONGRUENCE BETWEEN APPROPRIATION (USE) AND PROVISION RULES AND LOCAL CONDITIONS

Appropriation rules restricting time, place, technology or quantity of resource units are related to local conditions and to provision rules requiring labour, materials and/or money. In-built flexibility is needed to respond to changes in resource or economic environments.

*continued ...*

### DETECTION AND GRADUATED SANCTIONS

Users who violate operational rules are likely to receive graduated sanctions (depending on the seriousness of the offence) from other users, from officials accountable to these users, or both. The more noticeable cheating is, the better the chances of success.

### COLLECTIVE-CHOICE ARRANGEMENTS

Most individuals affected by operational rules can participate in modifying them.

### MONITORING

Monitors, who actively audit both the resource and the behaviour of users, are accountable to the users and may themselves be users.

### RELATIONSHIP BETWEEN USERS AND THE STATE

The ability of the state to penetrate to rural localities, and state tolerance of locally-based authorities is important: the less the state can, or wishes to, undermine locally-based authorities, and the less the state can enforce private (non-common) property rights effectively, the better the chances of success.

### CONFLICT RESOLUTION MECHANISMS

Users and their officials have rapid access to low-cost local arenas to resolve conflict among users or between users and officials.

### NESTED ENTERPRISES

Appropriation, provision, monitoring, enforcement, conflict resolution and governance activities are organised in multiple layers of nested enterprises (possibly linked with other democratically-based political institutions).

(Taken from Hobley and Shah (1996b). Sources: McKean, 1995; Ostrom, 1990; Wade, 1988.)

## Farmer service groups

Drawing a line between farmer service groups and common pool resource (CPR) management groups is not always easy. Groups which coalesce around CPR management may also provide services to their members to enhance the productive potential of the resources which they manage. This implies that farmer service groups and CPR groups are not always distinct. This section, however, focuses exclusively on the service-providing dimension of group activity, whether within groups formed expressly for this purpose or within groups having other, perhaps dominant, purposes. It will not, however, cover those groups which are effectively vehicles for NGO efforts in service provision.

### *What has been done?*

One of the most prominent and spontaneous examples of bottom-up service provision activities comes from the farmer-to-farmer movement extension in Latin America. This is based upon the principles that farmers' own knowledge and capacity for experimentation are a sound basis for the development of appropriate new technologies, and that farmers themselves are the best teachers since they can demonstrate new ideas being tried on their own fields to other farmers. As Holt-Gimenez (1996: 1) states, '[farmer-to-farmer] is not so much a programme or a project looking for peasant participation, but a broad-based movement with *campesinos* (peasants) as the main actors'.

The early proponents of farmer-to-farmer extension saw it as filling a gap in the Latin American popular education movement of the 1970s and 1980s, which was characterised by horizontal communication between learner and teacher, and by 'reflection-action-reflection' approaches for social transformation and local empowerment. The many thousands of social activist 'promoters' of this approach in Central America taught literacy, health care and community organisation but were weak in such technical subjects as agriculture. It was because of this that NGOs tended to become involved, and now the movement has developed and evolved into a farmer-led but NGO-facilitated approach (which is addressed in the following section).

More explicitly-defined farmer groups have been formed for a number of different purposes, ranging from the political to the purely commercial. On the whole, the strongest such groups are those which have links with the market or which have been formed expressly to cater to their members' marketing needs (although it should not be assumed that links with the market are themselves a sufficient condition for success, as the failure of many cooperatives in developing countries has demonstrated). Such groups have served their members through the provision of technical advice, dictated by the need to produce a particular quality of product for the market (for example, the El Ceibo federation of cocoa cooperatives in

Bolivia advises members on organic production techniques for sale of their product into the high value European markets). They have also provided their members with output marketing services, processing facilities, input supplies and even credit. Box 4 provides some examples from Uganda.

Formal, but not market-driven, membership organisations, such as the Zimbabwe Farmers' Union (ZFU) and the Uganda National Farmers' Association (UNFA), have also provided their members with valuable services. The ZFU, for example, has negotiated special credit terms and sales tax exemptions for its members, has assisted them with transport in remote areas and has enabled them to gain access to government warehousing facilities. UNFA has also negotiated certain product discounts for its members and has been building up its own extension service. Groups of farmers are expected to 'demand' training on certain topics and then to cover the costs of training provision. In the future, UNFA intends to broaden its activities further into marketing (it has already organised some bulking-up schemes), processing and credit provision with the assistance of DANIDA, the Danish government aid agency.

The proximity of large, federated farmers' unions to government structures (and, conversely, the genuine degree of independence that they enjoy) varies considerably. Some national unions, such as the Kenya National Farmers' Union and, to a lesser extent, the ZFU enjoy a fairly typical 'corporatist' relationship with government while others, such as the Malian Union of Cotton and Food Crop Producer (SYCOV) have challenged the authority of the government and related parastatals very significantly (SYCOV threatened to call a 'strike' of cotton producers and at one stage there appeared to be a threat that cotton mills would be set on fire) (Bingen *et al.*, 1995).

---

*Box 4* Marketing groups in Uganda

SOLAR FRUIT DRYING IN UGANDA

The Kyeirumba Women's Association and the Nakatundu Young Farmers' Group both supply the private trading company, the Fruits of the Nile, with sun-dried fruit for export to the UK. Initially, both centred their production activities around large group-owned driers which, in the case of the Kyeirumba group, was financed by an NGO. Over time, both groups have moved away from group production as they have encountered difficulties in managing the assets efficiently. In place of this, members have used their own savings to purchase small, individually operated driers. However, group arrangements have been retained for marketing, as the benefits (lower transport costs, regular deliveries to the trading company, payments by the company into a
*continued . . .*

---

group account at the bank) outweigh the costs involved. An indication of the success of both groups has been their ability to find a sustainable balance between individual and joint activities.

### UVAN LTD AND GROUPS OF ASSOCIATIONS OF VANILLA FARMERS IN UGANDA

UVAN Ltd is a private company which buys, processes and exports vanilla. It works with groups of farmers organised into associations. This facilitates the company's operations as the associations play a role in selecting suitable farmers for participation in the company's lending programme, recovering loans and bulking up and assessing the quality of vanilla for purchase. They also mobilise farmers to attend training programmes provided by the company and negotiate with the company on the price paid for vanilla each season.

The associations are registered under the register of companies. Each association agrees on common objectives and rules and UVAN Ltd provides some secretarial support and legal assistance. Each association has a voluntary executive committee and members are required to pay subscriptions. Benefits of association membership include improved access to extension services, timely marketing of produce and immediate cash payments, a collectively negotiated price for vanilla, and access to financial services.

(Stringfellow *et al.*,1997)

However, contrary to the beliefs of commentators such as Sims and Leonard (1990) and hopes of many others, formal farmers' organisations have had relatively little involvement in the development and transfer of agricultural technology. Although a number of organisations have made pronouncements about the importance of research and extension, there is relatively little evidence to show that these organisations can act as effective pressure groups and ensure that reluctant technology systems reorient their activities towards the needs of small farmers. Most formal farmer groups rely to some extent on members' financial contributions for their own organisational survival, and few are financially well-endowed. This means that they often opt to focus on activities such as marketing or input supply, which generate immediate and obvious benefits for members (with a knock-on effect on subscriptions) and, if possible, a revenue stream for the organisation (Carney, 1996a). An additional problem is that many members of such organisations are simply not aware of the potential benefits that new agricultural technology (such as improved seeds or more accurate pest control methods) can bring as they have had little or no contact with government technology services in the past. Such a problem is particularly acute in countries such as South Africa (Carney, 1996b). Lacking such awareness, small farmers are unable effectively to demand and draw down external assistance. In principle, the organisations of which they are members can help them to overcome this problem by introducing them to new ideas, but frequently these organisations are too resource-constrained and not sufficiently well-connected to existing agricultural technology services to be able to do this effectively.

### Outcomes and why

On the whole, small-scale farmer groups appear to have been more successful in meeting members' immediate needs than their larger counterparts. Indeed, one of the hypotheses of recent research conducted in this area is that the most effective primary groups have between eight and twenty-five members. Being small, they are better able than their large counterparts to ensure that their agenda is member-driven. Larger organisations wrestle with the problem of how to prioritise members' diverse interests and rarely seem able to sustain institutional coherence from the grassroots level upwards. This is a particular problem in the complex area of agricultural technology. For example, in the Marabá area of the Brazilian Amazon, farmers' unions are working in partnership with the Tocantins agricultural research centre to develop technology for the region. Researchers at the centre sometimes find the unions to be unduly influenced by voices from the areas in which union support is concentrated.[3] These areas are frequently unrepresentative of the broader region (Marcia Muchagata, personal communication).

In general, the larger farmers' unions have had most success not in operational activities such as technology generation or service provision, but in lobbying on issues such as prices, land reform and the maintenance of extractive reserves (in the case of the rubber tappers of Brazil (Richards, 1997b)). For example, by threatening that its members would 'strike' (that is, refuse to deliver cotton) the Malian Union of Cotton and Food Crop Producers (SYCOV) achieved significant restructuring in the Malian cotton system and secured a substantial price rise.

One reason why achievements in service delivery have been less notable is that service provision on a large scale, such as these organisations would need to operate to meet all their members' needs, is extremely complex (which is one of the reasons why the government itself has frequently failed to operate an effective system). Again in Mali, SYCOV has stated its desire to take over provision of agricultural inputs from the parastatal cotton company, but appears not to have given adequate thought to the requirements and difficulties that this would entail (Bingen *et al.*, 1995). Perhaps for this reason, the ZFU more often lobbies other bodies, such as the Zimbabwe Credit Board, to give its members preferential terms, rather than choosing to provide services itself. Economic feasibility studies are certainly not a strong point for farmer groups in general and while large unions frequently have elaborate formal structures in place, these are rarely functional when it comes to operational activities and financial accountability (Carney, 1996a; Stringfellow and McKone, 1996; Stringfellow *et al.*, 1996).

One notable exception to this 'rule' is the Uganda National Farmers' Association which has devoted substantial time and energy to building up its district branches (and the structures which lie 'below' these at village and district level). Under the tutelage of DANIDA, UNFA has also laid a strong emphasis on financial propriety and accountability, which now puts it in a strong position to expand the scope of its operations. This said, UNFA still lacks a certain amount of objectivity as regards its own internal capacity and ability to manage a very diversified portfolio of activities effectively. It also stands in danger of antagonising other players within Ugandan agriculture as it is perceived to be developing dominant tendencies.

### Preconditions for success

The findings of a recent review of the factors behind the success of farmer-controlled enterprise are presented in Box 5. The researchers in this study concluded that success cannot be firmly predicted in advance, nor can it be assured by an outside party; it depends in large part upon the motivation of group members and the congruence between members' skills and the challenges they face. For this reason, externally-driven groups often fail, even when some care has been put into devising appropriate group support strategies. The implication of this may be that donor investment is more likely to be productive when dedicated to general skill enhancing purposes (literacy, numeracy, etc.) than group formation *per se.*

---

*Box 5* Features of farmer groups usually associated with success

- Many of the internal features associated with successful group activity are inter-related. A *clear member-driven agenda* appears to be crucial, and while this is linked to participatory processes within the group, it is also likely to be linked to the size and internal cohesion of the group. Small, homogeneous groups have been shown to be able to operate these processes more effectively than large, heterogeneous groups. Equally, procedures to ensure
  *continued...*

---

accountability and financial transparency can be more effectively implemented by small groups. These findings are particularly relevant where group activity requires a commitment of financial resources to a shared enterprise. Where, on the other hand, the group's primary function is to liaise on behalf of its members with a buyer or supplier, larger and more heterogeneous groups may be able to operate more successfully.

- *Previous experience of group or cooperative activity* can make an important contribution to the development of cohesive groups. Women in particular have an established tradition of cooperative interaction underpinned by both social and economic factors.

- *The pursuit of a single activity* in the early stages of group development and achieving an effective balance between individual activity and group activity both appear to contribute to group success.

- A group must have *a strong business rationale* if it is to develop successfully. If a group forms simply to access material resources (whether on the encouragement of political patrons or in response to a disbursement-driven donor programme), its members will have no incentive to continue cooperating once these have been disbursed. An important exception to this rule lies with those groups which form to access working capital loans from financial intermediaries.

- *External training inputs* have played an important role in ensuring the success of many groups. Over and above the actual skills imparted to group members through such training, it is important to set the training within a broader economic framework, relevant to group activity.

(Stringfellow *et al.*, 1996)

If farmers' organisations are to become successfully involved in agricultural technology development and transfer, as opposed to direct service provision, they certainly need to have specialist skills. In particular, they must be able to engage with technology professionals on a relatively even footing to secure their own credibility. Often they lack familiarity with the agricultural technology system, to the extent that they do not even know where to focus their often very limited resources to achieve maximum impact. Box 6 details the major concerns, and suggests strategies for increasing the leverage of farmers' organisations in agricultural technology development.

*Box 6* Increasing capacity for technology involvement: areas for attention

IDENTIFYING AND PRIORITISING MEMBERS'
PRODUCTION PROBLEMS

If farmers' organisations are to maintain their legitimacy as representatives of large groups of farmers, they need to develop transparent priority-setting procedures. In large organisations, problem identification and priority setting is probably best done on a commodity-by-commodity basis, incorporating market information into the decision-making process where relevant. Special attention will need to be devoted to ensuring that the needs of the poorest members are not neglected.

INCREASING TECHNICAL CAPACITY

If farmers' organisations have reliable access to technically-trained personnel, their credibility in the technology area is likely to be enhanced. In-house agronomists can assist in problem identification, the search for likely solutions, training and sensitisation of members to the potential benefits of agricultural technology, and improving the quality of interactions with research and extension personnel. Where resource constraints prevent organisations from employing their own technical personnel, it may be important to encourage external resource people with technical skills to support them so that technical options can still be generated internally.

DEVELOPING RELATIONS WITH THE TECHNOLOGY
SYSTEM

A principal reason why farmers are unable to access technology is because they are unaware of how and where it is developed and who the key decision-makers are. Investment in understanding the system and developing cross-cutting webs of formal and informal relations over a long period of time is unlikely to be wasted. Ideally, organisations should be able to engage simultaneously at all stages of technology development. However, organisations of low-resource farmers are unlikely to have this capacity and will probably need to take a much more incremental approach. Donors, with 'insider knowledge' of the technology system should be able to assist in this area.

*continued* . . .

### INCREASING THEIR RESOURCE BASE AND IMPROVING FINANCIAL MANAGEMENT

Technology involvement tends to be expensive. Those organisations which are renowned for their 'success' in the technology area (such as the El Ceibo federation of cocoa cooperatives in Bolivia) tend not to represent the poorest farmers and have often received significant amounts of donor money. It is not sufficient simply to improve access to funds; organisations need to be able to manage that money efficiently. This is one problem which is often overlooked by enthusiastic, financially-constrained leaders. The Malian Union of Cotton and Food Crop Producers (SYCOV), for example, is lobbying to be able to access a percentage of its members' cotton payments directly. This would generate more than $1 million per annum for the organisation. Since SYCOV has neither a functioning office nor paid staff, it is unclear how this sum would be accounted for or how expenditure priorities would be developed within the organisation (Bingen *et al.*, 1995). Poor management and subsequent allegations of corruption can effectively destroy organisations as both potential donors and members themselves quickly become disaffected.

(Carney, 1996a)

Only a very few existing farmers' organisations in developing countries are adequately provisioned in the areas highlighted in the box. Those that are, however, such as the El Ceibo federation of cocoa cooperatives in Bolivia, have shown themselves to be very effective in meeting their members' technology needs though they usually require large amounts of external support to reach this position (Carney, 1996a; Bebbington *et al.*, 1996). El Ceibo has received in the region of $20 million over the last two decades and ZFU has received between 20–60 per cent of its annual funding from external sources. In Uganda, UNFA depends heavily on DANIDA to cover its capital costs and an annually declining proportion of its recurrent costs (declining from 100 per cent to 0 per cent over ten years). Donors and public sector agencies have also provided direct technology inputs and training; for example, the public sector research system in Bolivia supported the technology efforts of the regional organisation Coraca-Potosi, and El Ceibo's agronomist benefited from donor-financed overseas training. Both small and large organisations appear to have greater chances of success when they have heavy external inputs in training.

Otherwise, it is likely that the technology efforts of farmers' organisations will be relatively piecemeal and often take place in response to 'invitations' to become involved which are proffered by the public – or more rarely the private – sector. This interface issue will be more closely examined in Chapter 5. However, at this stage it is important to note that organisations which are successful in service provision are rarely completely independent. This is perhaps one key area in which service groups differ from CPR groups. Several traditional CPR groups have managed access to resources such as water and forests for years (Pant and Cox, 1996). Indeed, external intervention and the effect of alien pressures (such as those imposed by market exchange) have frequently been blamed for the downfall of such groups (Richards, 1997b).

## NGO approaches

In agriculture as in other sectors, NGOs have over the past two decades taken on an increasingly important role. Their position has been strengthened by the confluence of two trends. First, as part of the process of democratisation in many countries, they have been granted more space in which to operate. Second, they are in receipt of increasing amounts of donor funding; over 5 per cent of all development aid is now going to NGOs. Neither change is uncontroversial. NGOs have critics who argue, among other things, that they are more often accountable to external funders than to their own clients, that their approaches are not adequately self-critical, that they do not represent value for money, that they have inadequate technical expertise, and that the intensity of their efforts on a small scale raises unsustainable expectations and results in dependency in the long term. Larger NGOs are also criticised for their relations with government. Some say that they have become too close to governments which they once opposed, and hence that they have lost their capacity to act as independent advocates of the rights of poor people. Others argue that their very success at performing many of the activities which they have taken over is relieving the state of its rightful responsibility to its citizens. Certainly, a key unresolved question relates to the extent to which innovative NGO approaches to watershed management or the promotion of farmer-to-farmer extension, for example, can be replicated on a large scale within the public sector and what adaptations, both to the approaches, and to the organisational structure and performance of government, are likely to be required if this is to take place.

A further issue which NGOs are now obliged to address relates to their own roles and identity. Bebbington (1997) argues strongly that NGOs in Latin America are in a transitional phase from being organisations truly 'outside' government, indeed highly critical of government, to being partners – or perhaps even contractors – to government in a market-led development process. The view that he puts forward is that this leads to crises of both identity and legitimacy, which require fresh new approaches. NGOs must modernise if they are not to lose their best staff to the private sector. Perhaps the most effective way to face up to the new challenges is to transform themselves into 'social enterprises', market-based organisations with not wholly commercial objectives. In this way they might be able to guarantee their own financial survival and also to make more headway in helping to integrate the poorest of their clients into the ever more pervasive market economies.

## What has been done?

NGOs have been active in both resource management and service provision (primarily research, extension, seed multiplication and supply, and health care and storage systems for pastoralists). Their efforts range from the very small (a domestic NGO operating in a single area) to the very large. The Bangladesh Rural Advancement Committee (BRAC) has made loans to nearly one million rural people, and the Aga Khan Rural Support Programme (AKRSP) groups in Pakistan involve almost 100,000 people. A characteristic of NGO projects is that they seek to respond to the needs of women and of those who do not have secure access to land, the latter especially in areas of high population density. Their target groups are often, therefore, those groups of people who have traditionally been neglected by the public sector, though the extent to which they succeed in reaching the most disadvantaged groups appears to vary quite considerably (Farrington and Bebbington, 1993).

Being very diverse, NGOs function in a variety of different ways, making it difficult to generalise about their activities. Nonetheless, certain commonalities in modes of operation can be identified. Many NGOs operate through or with groups of local people. In some cases they engage in group formation in a fairly instrumental way. For instance, they may seek to exploit the scale advantages of groups by helping them to form around such immediately economic functions as the marketing of inputs or outputs. In other cases, however, group formation is a goal unto itself. For NGOs that operate in this mode, group formation is one important step in the process of empowerment of rural people. This might be a general aim or may be related to a specific purpose such as encouraging local participation in decision-making about watershed management (and the subsequent collective action necessary for implementation of decisions) (Shah, 1995; Ravnborg and Ashby, 1996).

Groups are often cemented through savings and loans schemes in which group members take collective responsibility for repayment by other members. There is a clear logic to the fact that such schemes and credit programmes should operate through groups. Groups effectively lower the cost of servicing borrowers and collective responsibility lowers the risk shouldered by the lender. It has been argued, however, that the most vital aspect of group lending schemes is the highly disciplined, indeed almost ritualistic, approach to making repayments (often in the form of compulsory savings contributed before a loan is even accessed). Members of Grameen Bank Groups in Bangladesh are obliged to salute, sit on the floor in rows and chant at weekly meetings and to treat Bank workers with considerable formality (Hashemi *et al.*, 1996). The benefits of group formation in savings and loans schemes appear, therefore, considerably to outweigh the ever-present difficulties of persuading people towards collective action (Mosse, 1996).

Other activities/approaches which are common to many NGO programmes are:

- participatory approaches to needs assessment;
- holistic approaches, addressing agricultural technology needs in the context of livelihoods as a whole (including supporting services to agriculture as well as health, nutrition and education);
- promotion of low external input, low-risk, environmentally benign technologies, developed and disseminated in such a way as to reinforce indigenous knowledge systems and local institutions.

It is not uncommon for NGOs to be drawn into management or service provision on the direct initiative of the government. This benefits the government both by reducing its direct costs (NGOs often access overseas funds) and by demonstrating its own commitment to reform. This is most commonly observed in Asia. In the Government of India's eighth five-year plan it is stated that 'the voluntary agencies which have abilities and demonstrate and innovate, provide technology and training and act as a support mechanism to the local institutions should be increasingly involved in the implementation of developmental programmes' (Shah, 1995: 17). Likewise, when the Philippines adopted its new constitution in 1987 it made a pledge to encourage NGOs which 'promote the welfare of the nation'. There are now several official forums within government ministries which deal with NGO issues, and particularly relations between them and government (Miclat-Teves and Lewis, 1993).

One area in which NGOs have been particularly active is the promotion of farmer-to-farmer extension. Although farmers themselves were the original catalysts behind the movement, it soon became apparent that it would have greater chances of success with the support (both technical and financial) of NGOs. NGOs have also been responsible for taking the principles of the movement – which developed as a response to new political, social and economic forces in parts of Central America in the 1970s and 1980s – to other parts of the world. They have combined these with their existing emphasis on participation to develop various highly regarded extension programmes (see ODI Agricultural Research and Extension Network papers 59a–c).

*Box 7* Establishing a farmer-to-farmer extension programme

Holt-Gimenez (1996) spells out the following steps in initiating and consolidating farmer-to-farmer extension processes:

1   Farmers who are willing and able to become 'promoters' are identified. Other farmers visit their fields. Diagnostic surveys are conducted.
2   Promoters lead workshops to identify ideas from other farmers and, where possible, from extensionists, for addressing the problems.
3   Farmers who will conduct experiments are selected, as are suitable sites. Options for experimentation are chosen.
4   Farmers visit and discuss experiments in progress.
5   Results are shared during visits by wider groups of farmers, and through farmer–experimenter workshops and media publicity. National and international cross-visits and farmer workshops are organised.

Promoters are a key element in process. Ideally, their resource endowments should not differ widely from those of other farmers, they should have an inclination and aptitude for experimentation and for sharing new ideas with others, and they should not be primarily concerned with monetary compensation. Some farmer-to-farmer programmes have developed a sophisticated hierarchy of specialisations among farmers. For instance, categories introduced by the Nepal Agro-Forestry Foundation include demonstrator farmers, farmer promoters and farmer trainers (Pandit, 1996). Some members of the latter category may eventually become employees of the supporting NGO.

According to Bunch (1996) the key principles of the approach have been: to motivate and teach farmers to experiment; to use rapid recognisable success rather than subsidies to motivate farmers to innovate further; to use technologies that rely primarily on inexpensive, locally-available resources; to keep the process focused by beginning with a limited number of technologies; and to train village leaders as extensionists. Box 7 shows the sequence of activities involved in establishing a programme.

In forestry, where the multi-agency or partnership approach to development seems to be more advanced, NGOs have usually acted in concert with government and groups of local people. Because of the publicity given to conservation issues in the forest sub-sector, and the fact that donors have favoured international NGO efforts at conservation over domestic govern-ment efforts, there are also a larger number of international NGOs, some of which have operated multi-million dollar, regional conservation schemes (Richards, 1995).

## Outcomes and why

NGO efforts in the agricultural sector have certainly received a good deal of attention and have been far-reaching in their effects. Bunch (1996) notes that from the development of the first principles of farmer-to-farmer extension in Central America in 1969, the approach and its derivatives have spread, mainly through NGOs, to some nineteen countries. Families' basic grain yields have tripled on a sustainable basis for a total programme input of less than US$400 per family. New technologies introduced include: physical soil conservation works; cover crops; composting; and intercropping with legumes. Holt-Gimenez (1996) notes that these have not merely contributed to yield increases; in addition, tens of thousands of hectares of degraded land have been rehabilitated in Central America alone.

Adaptations of the original approach have gained ground elsewhere where the spirit of voluntarism is strong and where agro-ecological condi-tions are not severely limiting. These conditions obtain, for example, in the Philippines (Baile, 1996), Indonesia (Sinaga and Wodicka, 1996), and Nepal (Pandit, 1996). However, two limitations must raise questions about how much further farmer-to-farmer extension in its original form can spread. One is its reliance on voluntarism; already a number of programmes have experimented with payments to farmer extensionists as a means of relieving this constraint (Pandit, 1996; Sinaga and Wodicka, 1996). The other is its reluctance to accept public sector research and extension as potential sources of appropriate technology. While the public sector may have had little to offer in Central America when the notion of farmer-to-farmer exten-sion first developed, this is not necessarily true at other times and in other places.[4]

Indeed, the benefits of forming links with various agencies have been demonstrated in a number of places, including the Indonesia Integrated Pest Management (IPM) farmer field schools. These have been established in over 15,000 villages in Indonesia and have received technical and organisa-tional support from FAO and from the Indonesian National IPM Training Programme (Kingsley and Musante, 1996). Another NGO initiative that has generated much interest comes from Zimbabwe where the Intermediate Technology Development Group (ITDG) has facilitated links between local people, the extension services (AGRITEX) and the Department of Research and Specialist Services as well as Zimbabwe Farmers' Union. Such modifica-tions to the original farmer-to-farmer approach are perhaps best termed 'farmer-led extension'. They have the potential to benefit from a wider range of technology options and offer the advantage of exposing public sector staff to new philosophies, so strengthening the potential for the public sector to incorporate these into its own operations. The future clearly lies more with these modified farmer-led approaches than with purist farmer-to-farmer models.

Farrington and Bebbington (1993) reviewed over seventy case studies of involvement by the larger NGOs in agriculture technology development. They conclude that NGOs have played an important role in technology development and transfer, despite severe weaknesses in technical capability in some cases. These weaknesses limit the type of projects which NGOs are able to take on. They militate towards joint approaches in which NGOs form linkages with agencies (such as government research and extension services) capable of providing technical inputs. However, the diversity of NGOs, in terms of objectives, philosophy and mode and scale of operation is very great. Some are more likely to form alliances with governments than others. Certainly, linkages may be difficult to cement with NGOs which were originally established as a forum for opposition to corrupt or authoritarian governments. Richards *et al.* (1996) give the example of the NGO CODEFORSA, which was set up with government support to work in the forestry sector in Costa Rica. Despite overall good relations with the government, it has at times suffered discrimination (delays in getting government permits etc.) because it has spoken out against corruption within the public sector. Bebbington (1997), however, points to a potential path which NGOs can follow in order to adapt to a new collegiate relationship with government. To some extent the proposed transition into a 'social enterprise' may compromise NGO relations with their primary clients (poor, rural people), but these are already strained by the 'self-appointed' nature of most NGOs as opposed to people's own elected bodies.

Overall, areas commonly identified by commentators in which NGOs have demonstrated quite serious shortcomings include:

- lack of ability or inclination to identify and address the processes and relationships underlying rural poverty;
- lack of ability or inclination to identify and exploit opportunities for influencing government policy;
- lack of ability or inclination to link groups with which they work into normal commercial channels for input supply and output marketing;
- lack of flexibility and innovations;
- poor pre-project appraisal and monitoring and evaluation;
- lack of evidence of cost effectiveness (in many cases the appropriate data simply is not collected).

(Farrington and Bebbington, 1993; ODI, 1996)

There is also consistent evidence to show that expectations of NGO capacities, especially in terms of their ability to reach the poorest, to target women and to incorporate participation into all stages of their projects, have been too high. Shah (1995) notes that NGOs working in the context of watershed management in Gujurat have so far operated on behalf of people; they have not yet empowered people sufficiently that they can make their own demands upon government. One reason for this is that some NGOs have a tendency to be somewhat paternalistic towards the groups which they create and to insulate them from the process of learning by trial and error.

### Preconditions for success

In many ways, the preconditions for successful NGO engagement in agricultural service delivery and natural resource management are the mirror image of the problems outlined above. The following are key considerations.

- *Greater clarity in relations between NGOs and local groups.* Since NGOs predominantly work with and through local membership groups, measures taken to enhance the quality of their relations with these have an overriding bearing on the success of NGOs' projects. The accountability of NGOs to group members needs to be established from the outset. It is also important for there to be transparency about the type, extent and duration of support that will be provided by the NGO.
- *Labour availability and income generation.* Labour is the chief asset of the rural poor. It cannot be provided in unlimited quantities to attend meetings concerned with consciousness-raising and group formation. If the poor can be expected to attend meetings at all, then they will do so only in anticipation of early concrete gains from new economic opportunities. Sooner rather than later, therefore, NGOs must take on the hard challenges of income generation for the rural poor. This may involve the creation of casual, unskilled employment opportunities through mechanisms quite different from the types of project in which NGOs currently engage.
- *Advocacy and the power to influence.* NGOs have tended to focus either on advocacy and campaigning, or on joint action in agriculture and resource management. An NGO wishing to draw on government technical services may jeopardise its position if it begins to campaign directly against government policy. However, if NGOs are to be successful they will have not only to influence government policy so that it moves in appropriate directions, but also to lobby for improvements in the execution of policy. There is important scope for combining resource management and campaigning activities within the same organisation or, where this may provoke hostility from government, for a division of labour between technically-oriented and advocacy NGOs, and collaboration between the two types. Bebbington (1997) does, however, point

out how difficult it can be for NGOs to cross-subsidise their advocacy type activities from the proceeds of more market-based activities (including any sub-contracting agreements they maintain with government). Though this may appear to be an attractive strategy, it is difficult to execute because in a competitive environment it will be hard to make 'excess' profits adequate to finance large-scale advocacy or poverty-focused activities.

- *Upgrading of technical skills.* Technical skills are important in agriculture and natural resource management; they cannot simply be relegated to second place behind social organising skills. NGOs need to build in-house capacity, or to learn how to draw in appropriate technical skills on an *ad hoc* basis. Like farmers' organisations, NGOs will find it hard to influence public sector technology policy and operational activities if they are not 'speaking the same language' as government staff. Many NGOs have already demonstrated a will to 'professionalise', both in terms of skills and modes of operation. This is also partly a response to pressure from donors.[5]

## Private commercial sector

This section focuses on the activities of the private commercial sector. Once again, it is hard to draw precise boundaries around this sector. Farmer service groups, of the type whose activities were described earlier in this chapter, may effectively act as private, commercially-oriented companies; El Ceibo in Bolivia and many farmers' organisations in China act very much as private companies, trading in the pursuit of profit. This section will, however, concentrate on private, commercial companies which do not have members (although they may have shareholders) and hence which do not primarily buy or sell from a group with which they have pre-existing contractual or semi-formal agreements of any type. It will not therefore address the question of outgrower schemes and contract farming.

### *What has been done?*

There have been two quite contrary approaches to increasing the activities of the private sector in natural resource management and service provision. The first has involved phased or one-step withdrawal by the public sector, allied to a hope or assumption that the private sector will step in to fill the resulting void. In some countries, policy makers have been able to build their strategies on existing evidence of private sector activity (for example, the existence of a thriving parallel market). In others, public sector retrenchment might be considered something of a gamble since very little evidence of private sector activity is available at the time of reform.

The second, rarer approach to increasing the involvement of the private sector has been for the public sector proactively to support the expansion of private commercial activity. This was what happened, after initial teething troubles, in the programme of reform of the public sector agricultural extension services in Chile, where the government provided vouchers to farmers to purchase services from the private sector on a sliding scale (Bebbington and Sotomayor, 1995).[6]

There are numerous examples of private sector involvement in all areas with which we are here concerned, including agricultural extension provision, agricultural research, forest exploitation (if not management) and the management of water rights or irrigation systems (Schwartz, 1994; Beynon and Duncan, 1996; Turral, 1995b). The exception lies in pastoral resources, where attempts to privatise rangeland management in sub-Saharan Africa have focused on granting rights to private individuals rather than intermediating private companies.

Other key areas of activity for the private sector include agricultural marketing, credit provision (through commercial banks), veterinary service provision, input and seed supply (see Carney 1995a for a review of experiences in all these areas). For example, the World Bank (1994) lists twenty-two African countries which, prior to reform, had both controlled markets and subsidised prices for fertiliser. Removing these subsidies and liberalising markets were easily identifiable goals for early structural adjustment loans, and by late 1992 only two out of these twenty-two countries are recorded as having achieved neither. Fourteen are judged to have succeeded in both while six of the original twenty-two retain some price controls. Similarly, Umali *et al.* (1992) show that, by 1991, in none of the twenty-nine African countries or nine Asian countries which they surveyed were veterinary services provided completely free of charge (although in some countries the public sector retained primary responsibility for supply).

It is now commonplace to analyse each of the various goods and services using the concepts of welfare economics (i.e. to classify goods as 'public', 'private', 'toll' or 'common pool'). This helps to predict the areas in which private sector involvement is likely to be economically attractive and therefore forthcoming. For example, it is generally held that the sale of hybrid seeds is a lucrative activity for private sector seed companies while the sale of open-pollinated seeds is not. The argument goes that research in open-pollinated seeds will not yield a good return on a large investment because users do not need to purchase new seeds every season; the benefits of investment in research on such seeds can therefore be appropriated by all, rather than just the investing company.

Many tasks which were formerly considered to be relatively unitary (and usually solely managed by the public sector) are more accurately and usefully regarded as complex packages with a number of divisible components. As an example, agricultural service delivery can be divided first into provision, financing and regulation. Each of the components of the package can be further sub-divided, as necessary; regulation can be divided into the laying down of the regulatory framework and its enforcement, and provision can be divided into setting the parameters of service, management of supply and actual supply. Once again this introduces a new level of complexity, but the advantage of such disaggregation is that governments can pinpoint exactly which activities they must retain and which can be performed equally well, if not better, by outsiders. This should help to guard against indiscriminate cost cutting, without regard to the effectiveness of services, which is rarely anything but counter-productive in the long term.

The concepts underlying welfare economics analysis are 'subtractability' and 'excludability'. If a good is excludable, those who have not paid for it are excluded from consuming it. If it is subtractable, one person's consumption of the good reduces its availability to others. These concepts greatly enhance our understanding of when and why the private sector is likely to become involved in any area. They are not, however, water-tight. Just as private companies can choose to base their marketing strategies on 'loss-leaders' (products priced at or below cost) to generate loyalty, so they can get involved in seemingly unattractive (in welfare economics terms) markets because of the size or overall importance of these markets. Thus Brenner (1991) argues that seed companies involve themselves in marketing open-pollinated seed varieties when they are building up a market position. They might remain in such markets if bulking up and packaging is a sufficiently onerous task that users do not wish to do it themselves. This is logical given that even if a small percentage of maize producers in a number of African countries chose to purchase their seed annually, the market is likely to be very large. Furthermore, research in the Great Lakes region of Africa has shown that it is the poorest bean producers who tend to purchase seeds annually because, unlike their richer counterparts, they cannot afford to save seed at the end of the season (Sperling et al., 1996). Similarly, in Kenya, Beynon and Mbogoh (1996) found that following liberalisation of seed markets in 1994, Kenya Breweries has been willing to invest in developing open-pollinated barley varieties. It recognises that the production of high quality barley is critical to its own commercial well-being. It also believes that since producers will have to repurchase seed every three to four years if they are to continue to benefit from its high yielding characteristics, the company will remain in a position to recoup its investment costs.

The implication of behaviour such as this (which seems to contravene welfare economics) is that guidelines about areas in which the private sector will become involved must also be set in the context of the particular markets and countries in question. The guidelines are indicators of the likely extent of private sector activity rather than firm predictions. Just as private, commercial companies can become more involved in certain areas than might have been deemed likely, so private sector activity can be notably lacking. This is a particular problem in the remoter areas of Africa where the economics of the provision of even fully private goods (such as fertiliser) are sufficiently bad, the risk sufficiently great, or the regulation of private trade still too onerous, that there is little private commercial activity.

### Outcomes and why

In principle, the private sector is much more effective (i.e., better able to meet clients' needs) than the public sector. This is because competitive pressures force it to respond directly to clients' needs or else risk going out of business. By contrast, public sector monopolies or quasi-monopolies are insulated from client pressures and the 'discipline' of the marketplace. An example is given of the contrast between the public sector and a private sector tobacco breeding programmes in Bangladesh. The private sector company identified the importance of smoking quality to tobacco acceptability, and tests of this therefore form an integral part of its breeding programme. The public sector has no testing facilities and, as a result, continues to produce varieties which have a low leaf quality, fetch low market prices and are rarely grown by farmers (Pray and Echeverria, 1991).

Srivastava and Jaffee (1993) note that the private sector is better than the public sector at providing packages of goods and information, effectively combining input supply and extension advice to gain a competitive advantage. Although this practice has the potential simultaneously to free the public sector of responsibility in two areas, it is not without its own hazards. The interests of farmers and of input supply companies are highly unlikely to be fully congruent, and concern is frequently voiced that partial and incomplete advice from company representatives is not in the farmers' best long-run interests. There is certainly a requirement for the public sector to retain the capacity to advise farmers on issues such as health and safety and how to get the same benefits using less not more of a particular purchased input (where possible).

In practice, however, the private sector has rarely been as successful in the agriculture sector of developing countries as might have been hoped. There are oft-cited examples of private sector involvement leading to enormous improvements in farmers' welfare; possibly the most dramatic change was seen in China when agricultural produce marketing was deregulated. There are also examples of unexpectedly high involvement by the private sector in areas which might not have been considered to be particularly attractive. For example, Pray and Echeverria (1991) report that nowhere in Africa does the private sector account for more than 10 per cent of agricultural research, but Beynon and Duncan (1996) found the proportion to be much higher in both Kenya and Zimbabwe.[7] However, accounts of failure of private sector companies or, perhaps even more common, of reluctance on the part of the private sector to fill the emerging gaps in the agriculture sector are far more common. Even in the areas which, in welfare economics terms, appear to be the most attractive to private companies, performance has been patchy. For example, both Kenya and Tanzania officially liberalised their fertiliser markets but public sector corporations continued to dominate (Mans, 1994; Swamy, 1994).

The root cause of this rests, it seems, in the inherent riskiness of commercial involvement in agriculture and natural resources more broadly. Indeed, the reasons used to justify public sector involvement in this area in the first place are proving to have perhaps more credibility than might have been assumed. Agricultural production is seasonal (which means that those purchasing or supplying inputs have to carry heavy inventories at certain times of the year, meaning that large sums of money are tied up). It is also highly dependent on seemingly random climatic factors. Especially in developing countries, producers are dispersed, produce quality is very uneven and demand for inputs and outputs varies considerably in tandem with the previous season's output and the general state of the economy. The lack of applicability of a number of these characteristics in the Chinese context – where private sector involvement is more apparent – is perhaps the exception that proves the rule. Farmers in China are closely spaced, the development of an effective road and communication infrastructure has been prioritised by the government and there is an almost insatiable market for agricultural produce in the country. Furthermore, farmers have become familiar with relatively high external input agriculture over an extended period of time.

As far as management of natural resources is concerned, the story is somewhat different. Frequently, private sector companies are *competing* with rural people for access to resources. This is nowhere more graphically illustrated than in the battles between private sector (or sometimes parastatal) logging companies and indigenous peoples in Latin America. In cases such as this, recent reforms have more frequently been aimed at substituting private management for management by local people (or at least creating forums to bring the two sides together) than at encouraging private sector activity. It should, however, be noted that there is a long way to go in this area and that immediate economic imperatives are frequently too attractive to governments for them to expel the private companies, whatever the longer-term environmental implications of private exploitation appear to be. Compromise solutions such as the establishment of joint decision-making forums which bring together private companies and rural people are rarely successful in that local people are operating from such a disadvantage that their views are seldom given adequate consideration.

One of the problems with private sector management of natural resources is that, if resources are to be managed with a view to the long term, investment costs can be extremely high. Thus while many forest product companies are happy to cut down virgin forest, far fewer are happy to invest in replanting for the second harvest. Likewise, it may be cheaper for those running large plantations to intensify production in the short term and then abandon land (and workers whose health has been damaged by inappropriate pesticide application, etc.). When it comes to water resources management, the costs of system building can be extremely high. As a consequence, private operators are often unwilling to become involved unless they can get significant up-front or ongoing public sector financial support. They also need to be convinced of the constancy of government policy and objectives.

The question of how government policy is perceived and how stable it ultimately proves to be is critical in both service provision and resource management. Many governments have emerged from decades of public sector control and then embark on reform programmes which are (to a greater or lesser extent) externally catalysed. It is rational for observers to question the commitment of these governments to reform, particularly when they can see that those who must execute the reform are often those who stand to lose the most (in terms of status, patronage opportunities and so on). *A priori* doubts are then borne out in a number of countries where the process of reform has been extremely uneven and new subsidies have been imposed after initial withdrawal, thereby reducing the attractiveness of private sector involvement.

In the absence of credible and attractive profit projections, there is no logic that will compel private companies to enter the market. Thus Sudanese efforts to introduce commercial private sector managers in its Gezira dam scheme have met with a very limited response due to concerns about stability and the direction of government agricultural policy, the quality of existing infrastructure, access to sufficient finance and taxation issues (Samad *et al.*, 1994). A similar reluctance on the part of private sector companies to take on contract management of irrigation schemes is reported from Colombia where lack of clear financial policy by the government is blamed (Ramirez, 1994).

In general, there appears to be very limited understanding of ways in which the public sector can support and nurture the nascent private sector without distorting markets unduly or creating expectations of financial support which it cannot sustain. Indeed, inadequate recognition has been given to the fact that reducing the direct involvement of the public sector in management and service provision does not necessarily reduce the overall costs which must be borne by the public sector. Since many reforms were enacted precisely to reduce costs, the possibility of effective government support on an 'infant industry' basis is frequently eliminated at the outset (Carney, 1995c).

Aside from the issue of the contrary signals which frequently emanate from government, major barriers to private sector involvement which seem to apply across most sectors are inadequate access to credit (this affects companies in two ways: they cannot gain access to start-up capital themselves and the buying power of their potential customers is reduced), inadequate physical infrastructure (the poor state of roads in many developing countries is often blamed for the weak penetration of the private sector in the more remote areas), and inadequate legal/accounting standards and consequently an inability to enforce contracts.

Additional, less widely applicable, constraints are inadequate access to foreign exchange (this is important where imports and exports are concerned and where financial deregulation has not preceded reform in the agriculture/natural resources sectors), past efforts to wipe out the trading classes, particularly when unpopular ethnic minorities have traditionally been responsible for trade (Bates, 1989), and problems of lack of commitment to programmes of reform in those skilled areas, such as the supply of veterinary services, where a limited pool of individuals must move over from the public to the private sector. Gros (1994) reports that in Cameroon, ministry officials had little faith in the reform programme and few skills which were relevant to its implementation. After four years, only twelve of the country's 112 veterinarians had set up in private practice and producer cooperatives had not emerged to take on any of the former public sector functions.

One controversy which has arisen in a number of countries concerns the role that multinational companies – most of which can circumvent many of the constraints mentioned above – should be permitted or encouraged to play. Some commentators and domestic decision-makers view such companies as representing the vanguard of a new form of imperialism, and reject them on that basis. Others welcome them for the investment capital and intellectual property which they can bring to previously isolated countries. A study by Echeverria cited in Dalrymple and Srivastava (1994) shows that tropical countries in which multinational corporations are conducting more research have a higher yield of maize (although the extent to which the yield differential is due to relative density of hybrids – hybrids being more attractive to the private sector than open-pollinated varieties – rather than the private sector *per se* was not addressed). It is areas such as seed research, in which investment costs are high but not entirely country-specific, which are the most attractive for multinational companies. However, such companies may need to team up with local counterparts which can provide market intelligence and control logistics and operations (this is, for example, the case in Turkey where seed market liberalisation is considered to have been a great success).

It should be noted that overall the agriculture sector has not proved highly attractive to international investors; the seed area is therefore something of an exception. Returns in agriculture tend to be quite low and risky, especially with commodity prices depressed as they have been over much of the past few decades. Additionally, investors are concerned about trade restrictions and imminent changes in trading agreements which might further restrict access to US/European markets (for example, the renegotiation of the EU's Lomé trading privileges and implementation of GATT/WTO Uruguay round agreements). Finally, there are few state guarantees available for agricultural investment in contrast to, for example, investment in infrastructure in developing countries.

Another overriding concern about private sector involvement as a whole is that it is likely to disadvantage the less well-off. This is the same problem that occurs when public sector cost recovery schemes are put in place; the focus is automatically on the relatively better-off clients. Since each purchase or sale made by/from such clients generates a relatively larger absolute amount, with a similar associated transaction cost, wealthier clients are considerably more attractive to profit-oriented enterprises than are poorer ones. In principle, groups of resource-poor farmers can come together to aggregate their economic power and hence overcome this problem. However, resource-poor farmers find it hard to create cohesive and effective purchasing groups as was mentioned in the section on farmer service groups.

That said, this problem should probably not be overemphasised. Certainly the private sector will tend not to reach the poorest, but then almost all suppliers find it hard to do so. One of the fundamental criticisms of the pre-reform public sector was that it was unable, or disinclined, to reach the poorest. There are numerous examples of this in all sub-sectors and with all actors. T&V extension systems were criticised for elitism, and public sector agricultural research for rainfed areas has been consistently underfunded by comparison with that directed at more favoured areas. The same trend can be seen in seed supply. Brenner (1991) reports that both public and private sector seed suppliers in Mexico and Brazil focused on the relatively more prosperous areas. Large farmers' organisations also tend to be dominated by the relatively better-off farmers (Carney, 1996a) and even NGOs have been criticised for their lack of poverty reach (Farrington and Bebbington, 1993). The larger question is then whether, as the private sector comes to cater to the needs of more of the relatively better off rural people, other suppliers (government, and non-profit organisations) can reorient better to meet the needs of the poorest.

Finally, the outcome of reforms to create space for private sector involvement in agricultural services provision may be hard to predict because of the complex and interlinked nature of many of the markets in question. When the seed market in Zambia was liberalised in 1990 and the Provincial Cooperative Unions there lost their position as monopsonistic purchasers, they also lost their access to credit. Instead of facing up to the new and potential beneficial competitive influences (they had tended to be very corrupt operations) they ended up unable to purchase seed at all. Some seed was sold through other bodies such as NGOs, but these were new to the business and not well organised, and general seed availability fell (Cromwell, 1992). Private traders failed to step forward because margins were perceived to be too low and they had little or no expertise in this area.

## Preconditions for success

It is beyond the scope of this book to address questions of the appropriate internal characteristics of private sector companies. However, these are not the only factors which are of importance. Successful private sector involvement in the supply of agricultural goods and services depends to a large extent on the overall 'enabling environment'. Suppliers will not be tempted to put large amounts of their own capital at risk unless they judge that the chances of success are relatively high.

The enabling environment is essentially controlled by the public sector, whose responsibility it is to lay down the regulatory frameworks for trading and the legal means to enforce contracts, to ensure relative currency stability, to impose reasonable levels of tax, and so on. The question of the interface between the public and private commercial sectors has of late been receiving increasing amounts of attention. While at one stage it might have appeared contradictory, it is now quite uncontroversial to suggest that 'for free markets to work better, the government must also work better' (Klitgaard, 1991) and that 'market-orientation and state minimalism, far from going together, are incompatible' (Streeten, 1996).

Analysts must, however, square up to the new challenge of mapping out the 'middle road'. The state versus market 'confrontation' (against which Sen (1996) warns) has given way to a debate about the 'extent and nature of market-friendly or supportive interventions' (Ranis, 1996). Concern centres on levels of funding, appropriate institutional arrangements and the details of the different roles of states themselves, of public agencies, private firms and households (Timmer, 1991).

It is expected that this line of thinking will develop further with the publication of the 1997 *World Development Report*, which refocuses attention on the public sector role in market-based economies. In the meantime, Box 8 details what might be thought of as the 'uncontroversial' public sector tasks in a market economy. These are classified as uncontroversial since most analysts would agree to them. Some, however, would wish to see the state taking on a considerably more interventionist role, especially in the agriculture sector. This might include involvement in areas such as ensuring price stability for basic foodstuffs and extending the direct supply of goods and services (for example, extension inputs) in poorer areas to improve 'social' as opposed to purely 'economic' efficiency.

In the area of natural resource management, relationships are more complex as the interests of different stakeholders diverge more widely. For private sector involvement to be considered a 'success', it therefore seems imperative that institutional mechanisms are established at the outset for consultation between all the different stakeholders. It is also likely that user groups will have to be well established and equipped for resolving matters

---

*Box 8* Uncontroversial public sector tasks in a market economy

It is widely agreed that to promote economic development it is appropriate for the public sector:

- to maintain and implement a consistent set of macroeconomic policies which are conducive to private sector development;
- to maintain law and order and an appropriate legal system with enforceable contracts and property rights;
- to provide an appropriate regulatory framework within which the private sector can operate in order to guard against the dangers posed by moral hazard and natural monopolies;
- to invest in infrastructure (both physical and institutional) to ensure that markets work better and information and goods travel efficiently;
- to invest in other public and merit goods (e.g., national security and education) and make efforts to internalise externalities, especially environmental ones;
- to make determined efforts to remove existing market distortions and privatise public corporations and parastatals (where appropriate).

---

internally before they can begin to engage with 'outside' commercial interests (Ravnborg and Ashby, 1996).

Finally, it should be noted that key benefits of private sector involvement (namely increased effectiveness and efficiency) are not in fact due to private sector involvement *per se*. Rather, they are benefits of liberalisation and the interplay of competitive forces. As Vickers and Yarrow (1988: 45) note, 'public ownership does not imply state monopoly, and private ownership does not entail competition'. However, they go on to say that 'although there is no logical connection between public enterprise and the absence of competition, there are several practical reasons why the two have often gone together'. These include the fact that public ownership is often the response to a situation in which competition is either not feasible or not desirable (for example, situations of natural monopoly or market failure); the undue influence of public sector managers on policy, and hence their ability to resist liberalisation; and the difficulty of maintaining a level playing field in situations of mixed 'public/private' sector competition. They conclude that the benefits of competition are more readily captured when the public sector withdraws entirely, although they also note that literature on mixed public/private markets is lacking.

## Conclusion

This chapter reviewed evidence relating to the roles of various non-governmental actors. Each different type of actor has different strengths and weaknesses. This in turn increases the potential benefits of exploiting complementarities by working together. However, it is important to note that almost all non-governmental actors (with the possible exception of the private commercial sector) have very limited reach. Common pool resource management and farmer service groups are both likely to function better when they have a clear and bounded number of members and the innovations for which NGOs have become known are often successful precisely because they are limited in extent (hence, they can be very resource intensive). Even commercial companies are likely to have to start at a small scale, unless they have the might of multinationals behind them. This makes the public sector particularly important in addressing concerns about equity and effectiveness with regard to the needs of the population as a whole (even taking into consideration the needs of future generations).

Individual farmers and resource users are clearly facing a much more diverse array of suppliers in the new environment (though the actual extent of options and now suppliers can vary considerably from country to country and even between areas within the same country). Given the diversity of users' needs, this can only be a positive development. However, the critical issue for the future will be to find a way to ensure that the interface between users and external service providers and managers functions in a way which is most likely to enhance effectiveness, efficiency and accountability. Users will certainly have neither the means nor the resources to interact on an intensive basis with all suppliers individually – just as suppliers cannot generally interact with all farmers on an individual basis. Identifying and strengthening appropriate forums in which such interactions can take place is likely to require considerable effort over the forthcoming decade.

*Table 2* Non-state approaches: a summary

| Principal types of change | Examples and outcomes | Future challenges for each type of change | Generic problems |
|---|---|---|---|
| 1 Changing roles of common pool resource management groups. | Forest management groups in Nepal and northern India:<br>• Some success, especially where (as in Nepal) rights and responsibilities are allocated according to local contexts or (as in India) markets for a particular product are good. | Wide range of tensions require close monitoring and management:<br>• Factors relating to internal management of group (size, composition, extent of coverage, nature of sanctions) of mutual obligations, and of mechanisms for conflict resolution.<br>• Factors linking the group to natural resources (nature and flow of benefits from the resource; procedures for deciding on rights and responsibilities and enforcing them; distribution of rights and responsibilities by income category of group members).<br>• Factors linking the group to state organisations (state tolerance of locally-based authorities; ability of the state to penetrate to local level). | • In remote areas, problem of linking members' demands with available resources.<br>• Problem for many groups of weaning themselves off support provided by e.g., an NGO.<br>• Groups are more easily formed around commercial commodities than around the production of subsistence crops.<br>• Without knowledge of members' needs, technical capacity to convert these into researchable problems, and knowledge of where/how key decisions on agricultural research priorities are made, it is difficult for groups to draw the public sector towards their requirements.<br>• NGOs may be more accountable to external funders; NGOs reluctant to allow early independence to membership groups. |
| 2 Changing role of farmer service groups (i.e., membership groups). | El Ceibo in Bolivia has provided marketing services for its 850 members for two decades, backing these up with the provision of inputs and crop husbandry advice. UNFA (Uganda) provides input supply, extension and marketing services to members. | • High cost of external support to organisations if they are to be successful.<br>• Lack of representation of poorest farmers. | |
| 3 The changing role of NGOs. | NGO support for farmer-to-farmer extension has permitted the wider adoption and stronger consolidation of the method than would have been achieved by links among farmers alone. | • NGOs mandate themselves to spend more time in a small number of villages than state organisations ever can, reducing the prospects of scaling up their approaches by government.<br>• NGO technical capabilities are often weak; they only rarely recruit the necessary skills or draw them down from government. | |

# 5

# THE INTERFACE

The evidence and conclusions presented in Chapters 3 and 4 point towards the potential offered by partnerships among different kinds of agencies (governmental, NGO, farmers' organisations, private commercial) in agriculture and natural resource management. Neither reforms within the state, nor approaches relying entirely on agencies outside the state are in themselves adequate to meet the expectations of a rapidly growing population seeking enhanced – or, at a minimum, stable – livelihoods from a threatened resource base.

The economic tools which were presented in the preceding chapter have already provided an indication of the appropriate division of labour between the state and the private commercial sector. However, our concern here is not so much with the division of labour, nor is it with those areas of service provision for relatively well-off farmers which can be quite readily devolved to the private sector. Rather, it is with the nature and scope of partnerships in which different types of agency plan and work together towards a common goal. Given the focus in this book on resource-poor farmers, the emphasis is more on partnerships between state and non-profit organisations (both service NGOs and membership organisations) than it is on partnerships with the private sector. Though there may be important opportunities for the public and private commercial sectors to work together to service the needs of the poor (for example in areas such as seed marketing and supply), ideas about such partnerships are only now emerging. They are not yet sufficiently developed (in terms of 'on the ground experience') to be covered in this book.

There is nothing new about the notion of state/NGO/membership group partnerships: they have been the rhetoric of numerous governments and donors over at least the past decade. Indeed, in some areas and sectors – such as primary health care in parts of South Asia – numerous practical examples of partnership can be found on the ground. These appear to work well. The experience in agriculture and natural resources has, however, been briefer and, so far, more problematic.

Programmes have often been embarked upon without adequate attention being given to preparation and analysis of existing capabilities. Government agencies have also shown themselves unwilling to act upon information which they have not generated themselves (Chowdhury and Gilbert, 1996). For their part, farmers, NGOs and farmer groups have not always taken into account the constraints under which public sector organisations function. These are manifested in the relative slowness of decision-making and inevitably more onerous bureaucratic procedures. The whole process has been compounded by the fact that neither the public nor the private sector has been consistently clear about its overall objectives and intentions in enacting reform and forming partnership (Turral, 1995b). Too often, mixed objectives lead to failed initiatives. Finally, the fact that policy conclusions and advice tend always to urge flexibility to match contextual conditions means that there are no blueprints available and the time investment required for partnership formation can be high.

Increasingly, however, donors appears to be becoming aware of these hazards. The Ford Foundation has recognised the importance of providing direct support to the government in India to create new partnerships (Hobley, 1995). This represents a partial solution to problems encountered in many areas, including in forest reform in Costa Rica, where inter-institutional agreements have faltered due to lack of resources. Sometimes, for example, new non-governmental partners cannot take on responsibilities unless they receive partial financing from the government during the early stages of implementation (Richards et al., 1996; Garg, 1995). Other donors are also investing more heavily in preparation and research into local conditions in preparation for partnership development. This is a critical issue, since the most effective policy changes are likely to be those which achieve coherence at local, regional, national and international levels (Richards, 1995).

Thus the rhetoric in support of multi-agency partnership is strong, but evidence about how such partnerships might best be designed remains sparse. The effectiveness of design will undoubtedly be enhanced as policy research over the next decade assembles evidence on the strengths and weaknesses of different approaches. Meanwhile, this chapter presents the preliminary findings of work sponsored by the Ford Foundation in which the ODI has been supporting multi-agency approaches at district level in Rajasthan, India. This yields some general guidelines and points to some important challenges for the future.

## Experience from Rajasthan

### Introduction

Udaipur district in southern Rajasthan exhibits many of the characteristics typical of low-resource, risk-prone farming. Sixty per cent of the cultivated area can only be cropped for the single monsoon season; at 18 kg/ha/yr, fertiliser consumption is lower than anywhere else in the state; topography is undulating and soils heavily eroded; 70 per cent of the cropped area is used for predominantly subsistence cereals (maize, millet); almost 50 per cent of farm families cultivate less than one hectare; and 55 per cent of the district's population of 1.8 million are tribal people.

As elsewhere, government extension has been characterised by centre-driven programmes to promote particular technologies, the most recent of which has been sprinkler irrigation. Such technologies are perceived by scientists and officials to be technically appropriate for the area and their promotion thus forms the basis for extension performance targets (number of demonstrations completed, etc.), while the provision of credit and subsidised inputs is linked to their use. The implication of this is that many villages have been subject to the frustrating experience of being offered little other than technologies which they cannot relate to their existing farming systems, and which they do not have the resources to support.

An additional difficulty is that village-level extension workers regard working in remote parts of this district as a 'punishment' posting. One esti-mate suggests that on average, 40 per cent of extension posts in the more difficult parts of the district are officially vacant at any one time while staff who have been engaged spend much of their time ostensibly sorting out 'problems' in the city. A recent recruitment drive produced a massive response; many of the candidates even had PhDs. However, the primary objective of most of those appointed appears to have been to gain a foothold in the public service. No sooner had they taken up their posts than many applied (and began to lobby through personal contacts) for transfer to more desirable locations.

### Ford Foundation involvement

Against this background, the Ford Foundation (FF) has since late 1992 been working with the Government of Rajasthan (GoR) to address the proce-dural and institutional issues which have constrained agricultural change. The programme aims to improve the client-orientation of agricultural services in Udaipur and at the same time to develop approaches in Rajasthan which have prospects of being implemented elsewhere. It is based upon collaboration between the GoR and local NGOs (Alsop, 1998).

At the time of the commencement of the programme, there had already been some experience in Rajasthan (and elsewhere in India) of the govern-ment *contracting* NGOs to provide services. The FF programme saw a number of limitations in this:

- the small scale of NGO operations and lack of scientific knowledge among NGO personnel limits the scope and spread of any productivity gain that NGOs might achieve;
- in at least some locations government staff are mandated to do precisely what NGOs are being contracted to do, resulting in wasteful duplication;
- contracts, which bind NGOs to the public sector as junior partners, limit NGOs' influence over the design and implementation of projects and procedures.

In addition, many NGOs saw contracting as an abrogation by government of its responsibilities. The underlying premise of the FF programme was, therefore, that interaction between the two sides should not be solely contractual, but rather collegiate. It should draw on NGOs' skills in participation and social organisation, and the technical capabilities, geographic coverage and funds of the government.

The principle of partnership is enshrined at national level in the Watershed Management Guidelines of 1994, and in the Guidelines for Joint Forest Management of 1990. It is also incorporated in GoR's Agricultural Development and Watershed Management projects, both supported by the World Bank. However, with a few exceptions, there has been reluctance on the part of Agriculture Department staff to implement this policy. The reasons for this include:

- the threat that they perceive to their own job security if NGOs are allowed to play a greater role in extension planning and execution;
- concerns (in some cases valid) over the financial probity of NGOs;
- concerns (again, in some cases justified) over the competence of NGOs to identify appropriate technical options with and for farmers;
- the inconvenience of having to respond to demands 'from the grass-roots';
- concerns over the levels of transparency and accountability to clients likely to be demanded by NGO partners.

FF's strategy proceeded as follows. First, it consulted a wide range of stakeholders to assess whether they perceived problems and potential solutions in similar ways. Next, it encouraged debate on how to move forward, and only once this was done did it design and implement some experimental action.

It has to be emphasised that this mode of open debate and search for solutions is unfamiliar to many in the Indian public sector, which is characterised by a strongly hierarchical command structure. Efforts to document and monitor the processes of debate, decision and experimentation therefore had to be introduced cautiously. However, emphasis was given to this process as it was seen as a potentially important aid to on-going learning and response. Following an unsuccessful search for an Indian organisation interested in and capable of managing this process, the Overseas Development Institute (ODI) in the UK was invited to do so, and has had a presence in Udaipur since mid-1994.

Initial discussions with senior and middle-level GoR officials, NGOs and farmers in late 1992 revealed a widely shared perception of the problem: government support to agricultural development was uneven in time and space, and of limited relevance to poorly-resourced farmers operating in rainfed areas. However, all parties were cautious about the prospects of developing close working links between government services, NGOs and farmers. The primary concern of the NGOs was that all decision-making power would remain with government, while the government was reluctant to form partnerships with NGOs which it perceived to have unacceptably low standards of financial accountability.

Both sides were, however, sufficiently interested to meet formally at state level to discuss the options, and FF circulated comprehensive minutes of the meeting. FF also provided a number of small grants in 1993 to NGOs interested in implementing joint projects with farmers' groups and government staff. Again, the results of these efforts were documented and circulated for discussion.

Important aspects of public sector culture came to bear at a second state-level meeting held in late 1993. While the organisational and procedural problems of working together were freely discussed, formal presentations and reports on activities made, at best, only oblique reference to them. In the words of the FF Programme Officer:

> Several reasons, which needed addressing if collaboration were to succeed, underlay this:
> - direct reporting can personalise issues and jeopardise relations;
> - these were not issues or topics which the actors were used to considering or felt they had any power to change;
> - no locally-controlled or institutionalised mechanisms for inter-organisational information exchange, debate or decision making existed at that point in time.
>
> ... It was apparent that activities could not be effectively undertaken, understood or modified unless there was information exchange, joint analysis and reflection and a degree of shared decision-making.
>
> (Alsop, 1998)

ODI was therefore requested to establish a process documentation and monitoring system and to stimulate intra- and inter-organisational learning and the development of decision-making skills. The range of techniques employed for this purpose is summarised in Table 3. ODI implemented these in collaboration with the Udaipur Farm Science Centre (KVK (see below)) which became an increasingly active partner over the months, eventually taking full responsibility for the best-known 'voice' of process documentation and monitoring in Udaipur, the biannual publication *Recent Developments in NGO–GO Collaboration*. This is now circulated in English and Hindi versions to some 150 NGO, government and donor officials within and beyond Rajasthan.

One of these process monitoring techniques – the quarterly forum – had an additional purpose, that of engaging NGOs and officials from research and extension services at state and district levels in debate and decision-making in ways as free from the rules of hierarchy as possible. All parties at the forum could exchange information on their current activities and future plans, invite each other to participate in or visit fieldwork, and begin to plan activities jointly. It also provided an important opportunity for NGOs to learn about each others' activities, and for government officials to become acquainted with NGO activity. The location of this forum was sensitive: it could not be seen as either a predominantly government-owned or predominantly NGO-owned activity. With the support of a FF grant, it was eventually located at a hybrid organisation: the Udaipur government Farm Science Centre (KVK), itself located at a prominent local NGO.

### New types, levels and processes of interaction

After two years of process monitoring and of meetings of the KVK forum, a number of outcomes have been achieved. These include both joint projects and a strengthening of the underlying processes of interaction. We first consider specific activities, and then processes.

### Joint activities

The opening of dialogue between NGOs and the Department of Agriculture (DoA) in Udaipur has resulted in an increase in extension services and materials available to farmers and farmer groups promoted by NGOs. For example, partly in response to the perception of an improved climate for collaboration, two NGOs initiated field activities in cooperation with the DoA during the 1994 monsoon season. The DoA decided to focus its demonstration efforts (including demonstration of short-duration varieties of maize, in which farmers had expressed considerable interest in the past) in the areas in which these NGOs were operating. For the DoA, the involvement of NGOs greatly simplified the task of identifying farmers to participate in demonstrations. This therefore facilitated the achievement of programme targets. Indeed, as of 1995, NGOs have undertaken forms of participatory planning with their farmer groups and have communicated requests for support directly to the DoA office. The DoA has shown itself willing to respond to these requests, often diverting resources in order to do so. NGOs have also linked with other agencies, for example, helping the farmer groups with which they work to secure access to institutional credit for agriculture (specifically through the National Bank for Agriculture and Rural Development (NABARD)).

Among the more innovative ideas for alternative extension approaches that have emerged from Udaipur is that of para-extension workers (PEWs). Proposals from NGOs in Udaipur prepared in consultation with the district office of the DoA and submitted to the World Bank's Agricultural Development Project (ADP) incorporated plans for the identification, training and deployment of PEWs. These individuals were to be drawn from the ranks of farmers in the villages in which the NGOs involved were already operating. Though these proposals have yet to receive support from the ADP they have been the subject of much discussion and debate.

An Agricultural Research Fund (ARF) was established under the terms of a grant from the Ford Foundation to the KVK. The ARF has sought to provide financial support to NGOs and farmer groups to commission research-related activities on their behalf. The ARF got off to a slow start, in part because of limited appreciation by NGOs of the benefits to be derived from improved access to researchers and sources of technologies. However, the KVK is addressing this problem by providing advisory assistance and training to NGO staff and representatives of rural communities in farmer participatory research approaches. One activity supported by ARF involved efforts to control ginger root rot. These were being undertaken by researchers from the Department of Plant Pathology of the university, in collaboration with an NGO and the DoA. The activity has shown considerable promise in developing seed treatment and biological control approaches which can be easily and effectively used by farmers.

There are now a number of projects and programmes implemented through government which seek greater participation by the private sector including NGOs, commercial firms, rural communities and individuals. These include the participatory development, privatisation of extension and contract research components of the ADP and the involvement of NGOs and rural communities in various watershed management programmes. Growing numbers of NGOs are involving themselves in a range of agricultural and natural resource management activities, supported by both the GoR and donor programmes. Government departments (such as Animal Husbandry, Horticulture and Water Resources/Soil Conservation) that lack networks of field staff see arrangements with NGOs as one of the few potentially viable means of reaching significant numbers of people and larger areas. Rural communities themselves are clearly interested in translating this surge in interest into improved services which will yield tangible benefits in the form of increased production and productivity.

## Processes

One of the main difficulties faced in Rajasthan has been that the resources available to government for supporting NGO–government collaboration are tied to specific funds and schemes, such as the Agricultural Development Project (ADP) and the Water Resources Development Project (WRDP), both supported by the World Bank. Early efforts by the FF Programme Officer to open dialogue between public sector agencies and NGOs at state and district levels led to the submission of proposals for collaborative projects by six NGOs in mid-1993. All six proposals were turned down because they did not fit the provisions of existing schemes, though the transfer of the GoR Secretary of Agriculture, who had been very supportive of joint activities, cannot have helped. Shortly afterwards, the Special Secretary for the ADP, also a strong supporter of collaboration, issued a press advertisement inviting bids for NGO support under the special initiatives component of the ADP. Again, however, of the ten received, most were screened out at an early stage because they did not meet the provisions of the ADP.

Following a second state-level meeting in February 1994, the Deputy Director (Extension) in Udaipur district, who was a strong supporter of collaboration, was given formal authority to make the resources of his office available for collaborative activities at field level, and to participate in the KVK forum and in ODI process monitoring activities. In late 1994 he invited proposals from private organisations, including NGOs, to assume responsibility for extension activities in specific areas. NGOs in Udaipur discussed this invitation in the KVK forum. They decided, however, not to apply on the grounds that the initiative represented an abrogation by government of its responsibilities.

Subsequent to this, and to avoid an impasse, new discussions were held with local NGOs. It was proposed that, where government posts were unfilled, representatives of farmers' groups with which NGOs were working would act as para-professional extensionists, taking over the duties (and some of the resources) allocated to the government extension workers. These para-professionals would be responsible for serving the needs of local farmers by drawing down information and technologies from the next level up in the government extension hierarchy.

However, these imaginative proposals (stimulated in part by the attendance of the Deputy Director (Extension) at a workshop on farmer-led extension[1]) were looked upon unfavourably by some because it was felt that they did not meet the specifications for extension provision set down by GoR. Naturally, NGOs found this excessively restrictive. A second reason

*Table 3* Characteristics of the main vehicles for process monitoring in inter-organisational collaboration in Rajasthan

| | Purpose | Mode | Frequency | For whom? | By whom? |
|---|---------|------|-----------|-----------|----------|
| 1 KVK 'Forum'. | Explore scope for interaction: <br>• actors get to know each other and existing activities <br>• some recording and monitoring of agreements via minutes, etc. <br>• 'satellite' interactions ('pairing off') | Verbal, but improved use of written media (agenda, minutes). | Quarterly. | Primarily NGOs and government organisations (GOs) *within* district. | Local organisations, *not* perceived to be either fully NGO or fully GO. |
| 2 Recent developments. | Vehicle for exchange of information on ongoing activities and for commentary (some of which aims to enhance accountability and transparency). | Written. | Six-monthly. | NGOs, GOs and international agencies at local, district, state, national and international levels. | Initially ODI, then local organisation as above. |
| 3 Village studies. | • understanding change at village/farmers' level <br>• understanding how farmer groups relate to NGOs and government services <br>• promote learning within organisations <br>• influence other organisations (GOs, funders) | Written output, but also a flow of internal discussions/ workshops. | Occasional written; continuous verbal dialogue. | Internal learning, but also informs NGOs, GOs and funders within/beyond district. | Initially ODI, now individual organisations (all NGO). |
| 4 Other studies. | In-depth reflection on selected themes (farmer participatory research, watersheds, methodology for process monitoring and documentation). | Written. | Occasional. | Mainly NGOs/GOs within district, but also beyond. | Some by ODI; others by individuals in GOs and NGOs. |
| 5 Process monitoring and documentation *within* organisations. | • reflections on own organisation's responses to linkages, pressures for change, etc. <br>• conflict resolution <br>• influence on future organisational policy, strategy, actions | Verbal/written. | Occasionally written; verbal may be ongoing. | The organisation concerned. | The organisation concerned. |

why they proved problematic was that the extension workers' trade union had become concerned about impending privatisation. The union threatened to call a strike if the proposals went ahead.

It is not yet clear how widely the GoR will implement partnerships in agriculture. However, what is clear is that efforts to document the positions taken by different agencies (see Table 3), the decisions taken, and the quality of interaction, and then to make this information widely available has begun to generate pressures within government for better performance. Greater consistency has also been demanded in the GoR's attitude to working with NGOs.

The difficulties faced by state-wide departments such as the DoA in moving coherently towards partnerships contrast with a number of less formal partnerships, some of which have 'spun-off' from the forum (Alsop *et al.*, forthcoming). These might involve, for instance, a village extension worker and a local NGO. The fact that such initiatives already exist suggests that partnerships can be made to work at local level where the potential gains are self-evident and each side can bring flexibility to the arrangement. However, for partnerships to succeed more widely in the DoA, a number of preconditions must be met.

### Preconditions for success

Experience in Rajasthan suggests that the following conditions must be met for the first steps towards collaboration in multi-agency approaches to be successful:

- A weaning of government officials away from the rigidities (but also security) of standard 'models' and procedures and the development of skills, self-confidence and procedures to cope with the idiosyncratic in positive ways.
- The overcoming of reluctance among line department staff to implement policy towards collaboration with NGOs in the spirit intended. In the Rajasthan case, imaginative approaches at very senior (e.g., Secretary for Agriculture) levels and much lower down the hierarchy (Deputy Director for Extension in Udaipur) were in danger of being bypassed or stifled by line department staff who lacked the necessary capacity for lateral thinking, or who felt their own positions threatened by the changes in hand.
- Human resource capabilities need to be built up on both the government and NGO sides. Government personnel require stronger inter-personal skills. They must understand better NGOs' objectives in forming groups and promoting joint action, and the difficulties they face. NGOs must develop stronger technical skills in agriculture among staff (or, in the case of smaller NGOs, they must be aware of where to access indepen-

dent advice quickly and effectively). Technical shortcomings make it difficult for NGO personnel to interact on the same level as government staff. In the Rajasthan case, they threatened to prevent the successful use of the Agricultural Research Fund which has been established to allow NGOs and farmers' groups to commission research from the public sector.

- Acceptance by both sides of the validity of documenting and discussing the *processes* by which joint actions succeed or fail and by which decisions are taken. Discussions and documentation can then be used as a basis for improved decision taking in the future.
- Establishment of mechanisms 'owned' by both sides which permit documentation in a non-threatening way (such as the *Recent Developments* publication) and which permit both sides to meet and engage in open discussion in which the influence of hierarchy is minimised (for example, the KVK forum).

Although progress in all these areas has been made in Rajasthan, complete fulfilment of these preconditions will clearly take some time. Both during the period of reform and thereafter, it will be necessary for government to screen NGOs' approaches in order to identify which can be implemented sustainably on a wider scale, and to assess what changes in the structure and procedures of government services are implied by each.

## Conclusion

The example from Rajasthan, as well as evidence from other areas in which research has been conducted, highlights a number of issues which are of critical importance if the interface between the various actors and users of natural resources is to be managed in the most productive way. In this conclusion we focus on three key issues and point to a further two which have thus far received inadequate attention.

### Linkage mechanisms

The Rajasthan case study has demonstrated the critical importance of finding the right forums for decision-making which can help to effect real change. Too often, reform is characterised by superficial changes. JFM committees such as those in Haryana (Chapter 3) replicate existing patterns of elite dominance, and inter-institutional consultation bodies are bypassed by those who take final responsibility for implementation. The design and strengthening of such forums is likely to have to be highly context-specific. However, critical factors for success are transparency and equality of access, as demonstrated in Rajasthan.

## Public sector responsibility

Despite much rhetoric about the increased responsibilities and leverage of the various types of non-governmental actors, the public sector retains a key role in implementing change. The public sector must show itself willing to embrace change if effective public–private partnerships are to be formed and multi-agency approaches are to function as desired. This may go so far as the state agreeing to finance the activities of its partners or at least agreeing to share its resources (for example, stumpage fees for forest exploitation) with local users.[2] As a consequence, donor funds which aim to promote reform are likely to have to be channelled not only to NGOs and farmers' groups but also to the public sector itself. NGOs and others may need to lobby for change in unresponsive public sectors, but this is likely to be a prior step in the reform process. It will have to precede the establishment of procedures for working together, because power imbalances usually mean that if the public sector remains reluctant to cooperate, it can derail or at least petrify any reform initiatives.

## The importance of combining technical and institutional approaches

While this research programme has not looked in detail at the technical components of NGO, farmers' group and public or private commercial sector approaches, it has highlighted the importance of technical expertise and of having an informed understanding of the issues at stake in the natural resource sector. Lack of meaningful technical content is blamed for the failure of many extension programmes and NGOs have been criticised for their often weak technical capacity. Likewise, farmers' organisations find it hard to engage with the public sector because of their lack of technical skills and non-governmental managers of irrigation systems have found themselves constantly reverting to public sector technicians and backup when problems occur. The importance of developing appropriate institutional solutions to take natural resource management forward should not, therefore, detract from further investments in deepening and broadening technical knowledge.

Two potentially important issues requiring further attention are:

## Scaling up small-scale approaches

NGO approaches to common pool resource management have exhibited different strengths and weaknesses from those of government. NGOs have used broadly Freireian techniques of conscientisation in a long process of face-to-face support for group formation in a small number of locations: group members develop awareness of the constraints they face, seek to address them from their own resources where possible, and draw on external sources (such as government) where necessary. This process typically graduates through small-scale savings and credit to the more complex issues of collective action in grazing or forest land management over a period of three to five years. By contrast, government support has to be spread over a wider area and even where technical measures are sound, group formation is generally weak. Key questions concern the extent to which government can replicate NGO approaches over a wider area, whether such approaches place unrealistic financial and organisational demands on government, and what other innovative approaches offer prospects for scaling up by government.

### Interface with political bodies

The research programme on which this book is based has largely focused on the dynamics of reform within the bureaucracy and in non-governmental organisations of one type or another. Although relations with political bodies were briefly touched upon in Chapter 3, this dimension was not addressed in any detail. Nonetheless, it is of critical importance: as farmers' groups gain in confidence, their ability to use representation on local political bodies as an alternative channel for influencing the work of line departments is a key additional asset. Furthermore, mechanisms for strengthening the accountability of executing agencies to representative bodies are ultimately the mechanisms which ensure that people's views are systematically represented in policy and its execution. It makes no sense to invest heavily in small-scale forums for 'participation' if the notion of participation and the right to exercise 'voice' is not carried forward into government more broadly.

The participation by farmers or farmers' groups in local political forums also poses a number of threats: factionalism along party political lines may split village groups and the role of NGOs – which may have helped to form such groups – may increasingly be called into question since, by contrast with elected political representatives, these NGOs are rarely clearly accountable to those they claim to serve. Despite these potential difficulties, in those countries where they are democratically elected, local political bodies offer rural people a powerful channel through which to voice their preferences, and there can be little doubt that they will grow in importance over the coming decades.

Table 4 Multi-agency approaches: a summary

| Principal types of change | Examples and outcomes | Future challenges for each type of change | Generic problems |
|---|---|---|---|
| 1 Joint Forest Management. | JFM initiatives in Nepal and northern India:<br><br>• Some success, but depends on willingness of state to foster local initiatives; depends also on how dependent forest departments are on revenue from JFM. | State needs to:<br><br>• Recognise long-established customary rights as well as formal legal rights.<br>• Promote investment in products other than timber.<br>• Allow groups to establish rules and sanctions according to local custom.<br>• Allow more flexibility in timing of lease payments by groups. | • Local groups need to strengthen links with local political organisations (e.g., *panchayats*) in order to enhance the accountability of state organisations.<br>• Legislative framework, e.g., as it affects local group formation in some countries, is prescriptive and does not permit flexibility and local adaptation.<br>• Performance monitoring within government departments is inadequate to detect and penalise underperformance in collaborative actions. |
| 2 Multi-agency approaches in agricultural research and extension (government departments, NGOs and farmer groups). | • Some success in Udaipur district in setting up mechanisms for closer interaction, for monitoring interaction through process techniques, and for designing joint actions that go beyond the contracting of NGOs by governments. | • Shared vision at district and state levels needs to permeate all levels of line departments.<br>• Administrative procedures at state level need to be made more flexible.<br>• Each side (government, NGO) needs to incorporate some of the skills of the other to strengthen communication (NGOs to take on some technical skills, government to expand interpersonal and mobilisation skills). | |

# 6

# SUMMARY

This book has been set against a background of increasing pressures on governments to enhance the relevance of policies towards the natural resources sector and the efficiency of their implementation. Most states have been pursuing varying combinations of three broad approaches:

1   focusing largely on improving their own capabilities in policy formulation, implementation and service provision;
2   ceding responsibility for certain services and management functions to private commercial or non-profit organisations, which in turn implies a more *laissez faire* policy context;
3   relying increasingly on alliances with private commercial or non-profit organisations which allow the comparative advantage of all partners to be exploited.

The book has examined evidence relating to the successes and failures of these approaches in four sub-sectors: agriculture (specifically, agricultural service provision), forestry, water resources and pastoral resources/rangeland management. It has drawn from them early lessons about preconditions for success in achieving improved effectiveness, efficiency and accountability to low-income users of natural resources operating in complex, diverse and risk-prone (CDR) conditions. In this summary chapter, we first review the new policy directions and then briefly apply these to draw some lessons at the sub-sectoral level.

## Reforming the state

### Scaling down of public sector efforts

Past government failure has led to intense pressure for public sector retrenchment. Where the agricultural economy is highly commercialised and physical infrastructure is good, there have been a number of successful cases of withdrawal by the state in favour of the private commercial sector.

However, even here there remains an important role for the state in the provision of 'public goods'. There are additional arguments for a continuing role for the state in more difficult areas: commercial organisations have been reluctant to enter areas where infrastructure is poor and demand fragmented, and the efforts of non-commercial alternatives such as NGOs and rural peoples' organisations in these areas have been patchy.

### *Enhancing the efficiency and accountability of the state*

Four broad directions can be identified among efforts to achieve this goal:

- decentralisation of decision-making authority and financial responsibility in the expectation of enhancing the relevance of services and the accountability of those providing them;
- enhancement of the capacity of the state to take on new regulatory functions;
- increase in the level of coordination across departments within the public sector; and
- increase in the level of regional cooperation between states.

However, the implementation of these changes has not been without difficulty. In a number of cases, public sector officials, fearing a reduction in their power base, have stalled the processes of reform. In others, the levels of skill within government organisations have not grown quickly enough to enable the full benefits of reform to be reaped or the preliminary impacts of reform to be monitored (so that positive lessons can be spread and corrective action taken where necessary). Current performance assessment criteria and reward systems need extensive redesign if they are to stimulate improved responsiveness by government departments to their clients. Overall increases in the sense of mission and purpose shared by public sector employees at all levels of the hierarchy, and their political masters, are also vital.

## Non-state alternatives

Involvement by the private commercial sector in CDR areas is limited. In some contexts, it may be possible to set in place the preconditions for private commercial entry but, on the whole, private voluntary organisations (whether involved in resource management or service provision) appear more likely than the commercial sector to offer alternatives to the state in CDR areas.

Increasing pressure on common pool resources together with misdirected government intervention have led to the demise of social arrangements for common resource management in many contexts. However, where the state or local NGOs have played a facilitating role there has been a resurgence of local membership groups. There is now optimism in some quarters that the future of sustainable resource management lies more with local people than with agencies of the state, though issues of elite domination and resource conflict may still be problematic.

There are, however, few examples of groups – whether formal or informal – established to provide technology-related services in agriculture, other than as a close adjunct to commercial activities. It is doubtful whether the few such groups that do exist can continue to do so without substantial external support.

NGOs undertake a wide range of activities, usually in support of local membership groups, in aspects of agriculture and natural resource management. In CDR areas, these have achieved a degree of success by enhancing productivity in environmentally and institutionally sustainable ways. However, most efforts are small scale and heavily resourced, raising doubts about whether they can be replicated by the public sector. Resourcing for NGOs is currently buoyant, offering the prospect of replication by NGOs themselves, but the growth in numbers of NGOs both willing and able to work effectively in CDR areas remains low. Further, in the absence of links with state organisations, some NGOs tend to be technically weak and are able to offer their clients only a limited range of options.

## Multi-agency partnerships

Rhetoric in support of partnerships between different types of organisations is strong, especially when it comes to meeting the needs of the poorest. The perception is that government agencies bring technical expertise and the resources for broad coverage, while NGOs and membership organisations bring skills in needs assessment and the organisation of joint action. It is anticipated that the effectiveness of the whole (working together) will be greater than the sum of all the parts (separate agencies working alone).

However, these approaches have proved difficult to implement in the natural resource sector. State agencies, especially in agriculture and forestry, have a long history of working as the sole agencies responsible for their respective sectors. They only reluctantly open the door to new partnerships, especially when the potential partners are organisations viewed by many in the public sector as lacking both technical competence and financial probity. NGOs, for their part, generally view the policies of the state, and the services deriving from them, as inappropriate to the needs of the rural poor. Many also consider the difficulties of achieving change in state policies to be intractable.

Most experience in this area has been achieved so far in joint forest management in south Asia, with particular success in settings in which the state has permitted locally adapted forms of access rights and sanctions to evolve, and where the support of local political authorities has been engaged. However, questions remain over the commitment of Forest Departments to these experiments, and over the degree of flexibility that they allow to local initiatives.

Irrigation management transfer aims to hand much of the responsibility for managing irrigation systems over to farmers. In practice, in most settings, the state has to retain important roles in water scheduling, maintenance and repair, and conflict resolution. Hence, the responsibilities are more accurately described as 'shared' than as 'handed over'.

Multi-agency partnerships in agriculture remain rare. Early experience suggests that commitment to these approaches at the highest levels of the state is still not shared by mid- and junior-level line department officials. Further requirements, if these approaches are to be successful, are that:

- government procedures for co-funding activities with NGOs must be made more flexible;
- each side will have to incorporate some of the skills of the other partner(s) if communication about objectives, process and outcomes is to be adequate;
- forums must be established to facilitate communication and collaborative activities between partners, to monitor progress and to make course corrections as necessary.

The analysis of *why* development initiatives have not achieved intended outcomes remains weak in the bureaucracies of many developing countries, as does the capacity to learn and spread lessons from previous experience. Unless this weakness is remedied, it will be difficult for the state to capitalise on the early lessons being generated by a diverse and experimental range of partnerships.

There are two distinct strategic tasks which must be confronted by governments and donors wishing to adopt this approach. They must:

- stimulate wide-ranging within-country interest in such approaches, through attracting political support and promoting and publicising the results and lessons of pilot initiatives;
- stimulate processes of learning so that institutional configurations, procedures and practices can be adapted to varying contexts.

Specific areas requiring support by government and donors include the following:

- Those working in the public sector must be given a new, task-oriented, sense of mission. As part of this, recruitment, performance assessment and reward systems in the public sector must be redesigned to stimulate collaborative action.

- Procedures within government for allocating funds and other types of support to NGOs and farmers' organisations tend to be inflexible. They must be streamlined and made more capable of accommodating the diverse requirements of multi-agency approaches.
- Necessary skills for closer collaboration (both with other agencies and with users of resources) must be introduced into public sector organisations. Key areas in which skills need to be developed are stakeholder analysis, needs assessment, management of the project cycle and conflict resolution. Closer collaboration does not come automatically, and it must be negotiated, not imposed, by government. Donors potentially have an important role to play in introducing and consolidating these skills.
- Monitoring mechanisms for new collaborative approaches must be put in place at the outset and procedures agreed upon for introducing course corrections as necessary.

Where multi-agency approaches have been attempted to date, the focus has been on interaction between outside agencies and government line departments. In some settings, local membership groups (sometimes with the support of NGOs) have also been seeking to make their requirements known through local political bodies. Isolated examples suggest that working through local political bodies offers an alternative, and potentially effective, means of making demands upon and enhancing the accountability of public services (as long as these political bodies maintain control over financial resources). Although long term in nature, the strengthening of democratic pluralism is a goal worthy of the sustained attention of both governments and donors. Multi-agency partnerships are facilitated in democratic societies.

Finally, it is clear that multi-agency approaches will be ill served if donor projects set up structures parallel to, but lying outside, those which already exist in the public sector. Donors must also take care not to establish procedures which are too costly to be maintained by local agencies when the donors themselves have withdrawn.

We now take each sub-sector in turn for the benefit of those with more narrowly defined interests. The role of donors is highlighted.

## Agricultural extension

A key policy question for the coming decade concerns the role of public sector village-level extension workers (VLWs), who form the lowest level of the public sector extension hierarchy. The Training and Visit (T&V) extension model pioneered by the World Bank sought to use these people purely to provide advice (rather than inputs, credit, etc.) to farmers. T&V also pressed for 'unified' extension services, arguing that advice on annual and perennial crops, horticulture and so on should be provided through a single system. This implied that VLWs should have well-developed skills in a number of areas. Such demands have now increased with calls for VLWs to learn more about systems interactions so that their advice becomes more relevant to individual farmers. They are also expected to act more as a two-way channel for information than in the past and to provide farmers with advice on matters such as credit availability and marketing prospects. The model then becomes one in which face-to-face advice is provided by well-trained individuals country-wide. This raises two sets of problems:

- the implied financial demands on the public sector are likely to be at odds with current efforts to reduce the role of the state;
- the expectations are particularly difficult to fulfil in CDR areas, where VLW posts require perhaps the highest skill levels yet are the least attractive to potential employees.

The scope exists for farmer-representatives to take on certain of the functions of VLWs. However, this is unlikely to be possible over a wide area unless ways are found of rewarding the new 'para-extensionists', either financially or through the provision of training and other opportunities. This is likely to cause tension and result in resistance from the public sector unless well-handled. Public sector/NGO resources are likely to have to be supplemented by direct contributions from farmers for the service they receive.

Para-extensionists will not be able to operate in isolation; their local knowledge should be supplemented with external technical advice if they are to fulfil their full potential. Such advice might come from neighbouring public sector VLWs or, more likely, from higher up in the extension hierarchy. In either case, changes are likely to be required in the skill levels of those responsible for providing advice. Ultimately, if para-extensionists are successful, the VLW cadre will be much reduced; this process needs to be handled sensitively to avoid blocking action by public sector staff so that the para-extensionists can provide appropriate information and communicate in an effective way.

As well as increasing the relevance and scope of face-to-face extension advice, attention should be devoted to developing innovative uses of mass media (radio, TV) in extension provision. This should provide a lower cost mechanism for reaching farmers.

Donors have an important role in providing technical and financial support to governments as they address these policy questions and design responses to them over the coming decade.

## Irrigation and water resources

Success stories in irrigation management by users' associations are generally based on small-scale or pilot efforts. There remain important questions over whether 'irrigation management transfer' approaches, which are now popular, can be successful in large, gravity-fed schemes. The preconditions for success here are that there should be:

- clear and sustainable water rights at individual and group level;
- compatibility between the irrigation infrastructure, user rights and local management capacity;
- clear responsibility and authority vested in the managing organisations;
- adequate financial and human resources for both the maintenance of physical infrastructure and the effective operation of managing organisations;
- transparent accountability of and supporting incentives for the managing entities;
- a reorientation of construction-minded irrigation departments to becoming agencies capable of servicing the needs of user groups;
- an arbitration service capable of settling disputes over the quality of service.

In the wider context, the conflict between urban and rural water use is growing. While sometimes it is agricultural water use that is highly subsidised, in other cases decision-makers put the needs of growing urban populations first. In areas around the major cities in some countries, this has begun to diminish capacity to produce food, generating higher demands on more distant sources of supply. Cases have been documented in which the transfer of water by administrative fiat is being challenged in the courts by farmers, underlining the need for allocations which recognise existing rights and which are founded on rational principles. Severe pollution in the environs of some cities has further reduced the value of water for other purposes. Major shortcomings exist in current mechanisms – whether legal, administrative or market-based – for resolving conflicts over access to water and water quality.

There must be regulatory agencies at river-basin level with sufficient powers to address these issues; in some cases, new structures will be needed to regulate inter-basin transfers. If, for convenience, the jurisdiction of such bodies needs to be restricted to existing administrative boundaries, then administrative units corresponding as closely as possible to natural watersheds should be chosen.

Donors can best support processes of change by:

- designing water assistance on the basis of country programmes, which set projects and programmes in the context of the key constraints to rational and sustainable water management;
- encouraging flexibility in arriving at mixes of economic, institutional and technological arrangements compatible with local contexts;
- supporting capacity building in both implementing and planning agencies and in such areas as the development of information systems, the conduct of public consultation, registration of water rights, contract management, demand management, cost recovery and monitoring, and regulation of water use and quality.

## Forestry

Research conducted in south Asia suggests that the privatisation of forest land is neither feasible (because of the high costs of protecting the resource) nor desirable (because of the potentially severe distributional effects, particularly on the resource-poor who depend heavily on the products of forests and wastelands). In this context, policies are best geared to the improved management of land that remains in public ownership through a variety of joint arrangements involving government, local groups and, on some occasions, NGOs. As with water resources, governments have primary responsibility for introducing many of the necessary changes, but donor-assisted training and pilot projects can support policy change relating to the following areas:

- *Redefinition of policy frameworks*: policy frameworks in several countries must shift from a restrictive to an enabling mode, not least in order to legitimise the support by forest staff of village resource management, and to allow local people to assert traditional rights to resources.
- *Support for local membership organisations*: effective support for local membership organisations can best come from joint approaches whereby combinations of NGOs, line department and local political bodies work together. Strong commitment throughout government departments in the context of an enabling policy framework is a precondition for such support.
- *Tradeoffs between environmental protection and poverty reduction*: these are likely to arise, especially when access to a forest area is restricted for a period. A strong policy response, which recognises this tension and which is developed in consultation with local people, is essential to the resolution of conflict.
- *Recognising the value of non-timber forest products*: forest departments must be stimulated to recognise the value of forest products other than timber and to accommodate the needs of the users of non-timber forest

products in their planning. If local management is to be successful, a protected resource has to yield both short-term and long-term benefits.

## Rangeland

Settled ranching systems, drawing on models from the USA and Australia, formed the focus of policy support for rangelands in many developing countries until the 1980s. However, under sparse and variable rainfall, such systems are able to support much lower herd densities than pastoral systems, which have the flexibility to follow rainfall patterns and which are therefore better able to accommodate risk. For governments to support pastoralism, a policy shift from rangeland *conservation* to sustainable rangeland *production* will be required.

Donor projects in the 1980s typically aimed to provide services to pastoralists, including marketing support to stimulate offtake and so reduce what were perceived to be excessively large herds. Now that the benefits of pastoral systems are more widely perceived, marketing support is best restricted to years of drought when it is necessary to provide herders with an income for cattle which would otherwise perish.

Partly because of the prevalence of inappropriate models for rangeland management, and partly due to a lack of resources which prevented the state from maintaining an adequate presence in remote areas, the public sector has gained only a limited understanding of traditional institutions for rangeland management. It has therefore been unable to focus its resources in ways consistent with local practice and tradition. In particular, it has been unable to take traditional institutional arrangements into account in the design of conflict resolution mechanisms or of legislation governing land tenure. Changes in access rights to land that occur as a result of human and livestock population pressure are likely to remain the subject of dispute unless the state plays an adequate facilitating and regulatory role.

In some countries, quasi-legal enclosure has been encouraged by partial legislative and land tenure changes. Though reforms remain incomplete, they have assisted the breakdown of customary grazing rights. Traditional leaders have allowed local elites to fence off grazing land so that the majority of small-scale herders are being forced into shrinking communal lands. Policy reforms are needed to regulate enclosure more carefully and in consonance with local socio-economic and agro-ecological differences. Care must be taken to ensure that reforms are accessible to the majority of poorer people and that longer-term investment is not stymied during the years of pre-reform preparation and subsequent transition. It is in this sphere of changing land tenure arrangements that support from donors to identify appropriate approaches may have the highest payoff.

# 7

# CONCLUSIONS

This book has brought together the findings of three years of research conducted in three continents (Asia, Africa and Latin America) and relating to four sub-sectors in the agriculture/natural resources area. Many of the conclusions it has drawn are therefore broad in scope, and those who are seeking more detailed analyses must refer to the source papers which are cited in the text. However, the fact remains that evidence from very different areas frequently points in the same direction and that lessons learned in one sub-sector can subsequently be applied to another. Since the emphasis for the future is on increased responsiveness, adaptation and learning, it makes no sense to base learning on only a narrow sample of experiences. Certainly users of resources interpret their needs in the context of their overall livelihoods, rather than in a manner which fits in with existing highly segregated government departments.

## What are the achievements to date?

The management of natural resources, and the supply of those services which enable the resources to be used to their best effect, is a dynamic process in which change is always to be expected, in line with the evolving needs and capabilities of users. Nevertheless, the change that has been witnessed over the past few decades is unique in extent. It is not just the way that the resources are managed that has been altered; it is also the underlying thinking behind the principles of management which has changed. Within this process, the rights of users have altered but so too have their responsibilities. Users are now expected to participate to a far greater extent, to take centre stage in debates about resource management, to exercise their demand (whether individually, through membership organisations or through market mechanisms) and to fulfil their maintenance and management obligations with a minimum of outside assistance.

At the same time, enormous institutional changes have taken place. Most importantly, a large number of different types of organisation have come

97

forth to represent and service the needs of poorer farmers. In some cases, new organisations have been formed within both the commercial and non-commercial wings of the private sector. Some of these are even promoted by the public sector (through extension groups, for example). In other cases it is more a question of pre-existing groups gaining the legitimacy which permits them to come forward and interact with outsiders. Still such groups are emerging, and still outsiders have only very limited knowledge of why these groups form, what binds them together and what kind of support they require if they are to make a full contribution to the process of development in rural areas.

The goals of reform are varied. The three key concepts which were introduced in Chapter 1 of this book are effectiveness, efficiency and accountability, and it can be argued that it is against the achievement of these targets that reforms should be judged. It is our contention that progress has been made in all three of these areas, but unevenly from goal to goal, from resource to resource and from country to country. The rhetoric of participation and accountability has certainly permeated far and 'beneficiaries' of programmes are, slowly but surely, turning into 'stakeholders' and clients. In the World Bank's new rural development strategy, for example, consultation and consensus building are given new prominence within a pluralistic vision of the way forward. Yet there is still a long way to go. We tend to hear about the success stories, the areas in which change has taken place, and not the residual areas where improvements have yet to be witnessed. This makes it hard to form opinions about the overall state of reform and the extent of the challenge which remains to be faced.

The stringent requirements of efficiency (effectiveness at lowest possible cost), make it difficult to draw conclusions at a general level about whether this goal has been met. However, one reason why research of the type reported here is conducted is to help improve the efficiency of reform by drawing out and disseminating lessons and guidelines from early reform efforts, so it is hoped that some progress is being made. It would, however, be most revealing to make a concerted study of the relative cost effectiveness of service provision and resource management over the period of the last decade. Despite elevated expectations in some quarters, public sector budgets have not generally fallen, except where resources have simply run out. Reforms have led to some improvements in effectiveness which should, therefore, signal an overall increase in efficiency (if budgets have not risen). However, the activities and spending of other partners in rural development must be taken into consideration if overall judgments about efficiency are to be made; many NGOs, for example, have been widely criticised for their lack of cost effectiveness and budgetary control. Their tendency is often to put large amounts of resources into intensive activities in a single area. The results on the ground can be impressive but systemic cost effectiveness may be jeopardised.

Chapter 1 also introduced the notion of 'social efficiency', a loose concept which embraces the idea that the state *should* intervene in market allocations where the social outcomes of not intervening are unacceptable. This is one area in which there appears to have been relatively little progress. States have not yet shown themselves to be capable of promoting, facilitating and assessing markets and then selectively intervening to further the interests of those who are excluded from market transactions by virtue of their poverty or location (or both). Public servants are still, it seems, more likely to engage with relatively better-off rural people, their traditional clients with whom they have some experience, than with the poorest of the poor whose needs are so much harder to identify and to satisfy.

### The way forward

Multi-agency partnerships, the option for progress which is advocated in this book, will not provide miracle solutions to the problems of rural development or to the challenges inherent in achieving poverty reduction in rural areas. They do, however, represent a constructive approach to the tasks in hand. The idea is that all the various organisations working in rural areas should cooperate where and when this is beneficial. They should talk to each other, learn from each other and contribute to ensuring that shared goals are met. This does not, however, mean that they will meld together to form a single, monolithic institution with a single, paternalistic notion of development and how the needs of rural people are to be met. Such a fear has repeatedly been expressed in NGO circles when NGOs themselves have cooperated with government and participated (as contractors or partners) in government-led programmes of development. This fear is well-founded to the extent that a monopolistic development authority is never likely to meet the highly varied needs of a highly varied population in an effective way.

However, early – and admittedly quite fragmentary – evidence relating to multi-agency partnerships suggests that there is scope for partnerships to be productive but not all-embracing. Multiple points of interaction and various different forums for consultation between different parties can be identified or established. Each one may need its own 'rules and regulations' so that the terms upon which agencies come together may be different according to the problem to be addressed. Such rules may be specified within the funds which support the interactions or may be hammered out on a case by case basis. Certainly, this variability means that adequate resources, time and personal commitment must be invested in ensuring that the partnerships are meaningful and viable over the longer term. New processes and ways of working together are unlikely to evolve without the investment of dedicated effort and a shared sense of mission, possibly to be stimulated by changed personal incentives (whether monetary or otherwise) to work together in the interests of poorer clients.

Multi-agency partnerships seem, then, to provide a promising way forward on the institutional front. It is, however, important to establish where the balance between institutional innovation and investment in technical change should lie. This can vary considerably according to the resource in question. In the water resources area, for example, investment in both the institutions which manage water supply, irrigation, sewerage and drainage and in the technical aspects of water management are likely to be required if future problems of water shortage are to remain manageable. Likewise, NGOs working in community extension are likely to have to engage both in community mobilisation and in the introduction of new technologies into the area, though the immediate emphasis may well be on capacity building (later, partnerships with the public sector can be used to draw in greater technical content). When it comes to rangeland resources, however, resolving issues relating to tenure and local management and the rule of law may be greater priorities than conducting technical research. There will also be variation according to the physical environment in which work is taking place. In CDR areas greater investment in local capacity building as a means to generate knowledge-intensive solutions to local problems may be most important while in more favoured areas in which producers are already highly integrated with the market, finding solutions to technical constraints (e.g. through improved input use) may be the priority.

## What are the main challenges for the future?

There remain a number of generic issues or challenges which underlie the continuing process of reform and the definition of new relationships between the state and the individual in all sub-sectors considered here. Certainly the new, pluralistic environment is not without its problems; these must be taken into consideration when planning reform or analysing lessons from existing programmes of reform.

### *Rural poverty alleviation*

The incidence of different types of membership groups, NGOs and private commercial organisations varies enormously. In some places there is probably needless replication of tasks, while in others there are large numbers of rural people who are neither represented nor serviced by any type of organisation. The problem of ensuring that the poorest, living in complex, diverse and risk-prone areas, are adequately participating, are well represented and enjoy beneficial links with ever more dominant markets remains very pressing. Decentralisation within the public sector is supposed to bring services closer to these people and to ensure that their needs are recognised and attended to, but decentralisation does not automatically resolve problems of elite domination. Government that is, in principle, more accessible to the poorest because of its proximity is also more accessible, and often more often accessed, by the richer members of society. This remains a challenge which must be directly addressed. Greater targeting of the poorest is likely to be required and evaluations of the specific effects of reform and project interventions on the poorest must be undertaken to help guide action in the future.

## The paradoxes of the public sector

Public sector organisations have faced, and continue to face considerable criticism. The new emphasis on accountability, efficiency and transparency demands thoroughgoing reform within many organisations which have hitherto epitomised the antithesis of these concepts. The progress of reform has nevertheless been slow and halting, and results of donor-sponsored efforts to promote civil service reform have generally been disappointing, often because they are too ambitious and inadequately specified at the outset (OECD/DAC, 1997). Beyond this perhaps soluble problem there are two more fundamental constraints which thwart improvements in the capacity of the public sector to stimulate, facilitate and regulate pluralistic, market-driven economies.

The first, termed the 'paradox of power' by Rondinelli, concerns the fact that for reforms to be successful there must be widespread political support and participation, yet such a change is viewed as a threat by those in power. Unless these people are committed to the notion of reform and willing to cede some responsibilities, success is unlikely to be forthcoming (Rondinelli, 1993; OECD/DAC, 1997). Where programmes are embarked upon in a defensive mode, so that movement towards reform is viewed as a 'concession' which is only acceptable so long as it serves to safeguard the stability of underlying power relations, the real goals of the reform are seldom achieved. This 'paradox' can be the cause of delaying tactics, seemingly *ad hoc* and unfocused reform programmes and reform which remains incomplete, often because financial resources to back up bold policy pronouncements or decentralisation measures are deliberately withheld.

The second constraint, termed the 'orthodox paradox' relates to the apparently contradictory pronouncements about the scope of state activity and the size and resources to be allocated to the state (Grindle and Hilderbrand, 1995). A minimalist state is expected to be able to see reform through and regulate and facilitate a new and unfamiliar environment. This implies that a state which was characterised by very poor capacity pre-reform is expected suddenly to develop the capacity to undertake far more complex tasks. Clearly this is an unrealistic expectation and one which must be tempered by longer-term expectations and probably a more resource-intensive process of change.

## Defining the degree of partnership

The strong conclusion of this book is that only multi-agency approaches are likely to prove adequate to meet the growing challenges in natural resource management and agricultural service supply. We do not, however, go so far as to say that all activities of all the organisations involved in such partnerships should be entirely congruent. The strength of all these organisations, on which the multi-agency approach seeks to build, derives in part from their ability to take independent decisions and to develop independent positions from which they are then prepared to negotiate. NGOs should retain the capacity to criticise the government as well as to work with it. Membership groups sponsored by NGOs (or others) should ultimately reach a position of independence from those NGOs; indeed, this might be seen as the final fulfilment of NGO empowerment efforts. The public sector must retain a certain distance between itself and the private sector if it is to be an effective regulator in later times.

The difficulty then comes in defining boundaries to activities: when should cooperation begin and opposition end, and how should resources be allocated internally to the different activities which all the various bodies undertake? These are not, of course, questions to which conclusive answers can be provided at the generic level. However, they are questions which will need to be addressed and managers and politicians should actively be seeking experience in this area to enable them to make more effective decisions on a case-by-case basis in the future.

## Finding the appropriate balance of power

The discussion of multi-agency partnerships in Rajasthan (see Chapter 5) emphasises the negative effects of fear and mistrust on partnership formation. Sometimes such sentiments are well-founded, in which case additional care is required in designing the parameters of a partnership and in addressing the underlying concerns before they are able to manifest themselves in damaged relations. Sometimes, though, the fears are based on misinformation and apocryphal stories; they can also be very personalised. One way of reassuring potential partners who are beset by such fears is to define in quite specific terms at the outset where the balance of power in any particular partnership will lie. This is, however, easier said than done, particularly since the exact composition of individuals participating in collaborative work can – and usually does – alter over time.

A fundamental concern relates to the power of the public sector within any partnership. The state certainly occupies a unique position and, too often in the past, this has led to abuse and overly dominant behaviour. Equally, though, the state must operate according to broad policy guidelines established – in democratic societies – by its political wing. It cannot put resources which it has gathered in taxes from others *entirely* at the disposal of local people with no accompanying guidance or regulation (not least because of the short-termism which is inherent in much private sector decision-taking, whether commercial or otherwise). There arises, then, a question of balance and potentially conflicting agendas. Similar, though not as intense, difficulties arise when NGOs take on some of the functions of the state and interact directly with local groups. The tendency is still for local people to be spoken for, rather than for them to speak for themselves, with the balance of power being tilted too far in favour of the NGOs with no third-party arbiter to point this out. It is unclear what safeguards can be inserted into partnerships so that these problems can be resolved before partnerships decline, become irreparable and are abandoned.

Unless sufficient publicity is generated about positive instances of 'power sharing' and sufficient attention is devoted to understanding the modalities which govern the partnerships in question, stories of negative experiences may well kill the whole notion of partnership. Again, careful exchange and information analysis are likely to be critical.

### The role of donors

Specific recommendations about the role of donors in each of the sub-sectors with which we have been concerned are presented in Chapter 6. However, given the current climate of 'aid fatigue' and the questions that are being raised about the role and effectiveness of development aid in general, there is a requirement for these recommendations to be placed in some context and for the continuing problems – and challenges – associated with development aid to be recognised.

Official support to agriculture and natural resources has been steadily declining over the past decade (though UK aid has remained quite steady, at least in nominal terms). This decline has taken place despite widespread recognition of the need for further investment in the agriculture sector. Reasons for the decline include the reduction in overall aid spending by some of the larger traditional donors, the relative lack of success of past agriculture sector projects, lobbying by developed-country agricultural interests; disillusionment with the type of large-scale infrastructural investments (e.g., irrigation systems) which formerly accounted for high proportions of natural resources spending, and concern over intensive agricultural practices which have been promoted by some donor projects (Carney, 1997).

At a general level, several studies have been produced which question the effectiveness of aid spending. For example, Burnside and Dollar (1996) show that there is no general correlation between levels of official development assistance received and a country's economic performance. Official development assistance (ODA) is associated with improved economic performance when it is provided to countries which are already implementing good macroeconomic policies. But since disbursement decisions are not made on this basis alone, the aggregate success of ODA in achieving this goal has been substantially reduced. As regards poverty alleviation, Killick (1997) contends that there is little to suggest that ODA has had clear poverty-reducing effects. Indeed, it has been shown that if countries become 'chronically aid dependent' ODA will have little prospect of achieving either of these desirable outcomes. At certain volumes and under certain conditions, aid may actively hamper a country's prospects of achieving self-sustaining development (SIDA, 1996).

Nevertheless, we believe that continued aid spending in the natural resources area is important. It was argued in Chapter 3 that state minimalism – and by association large budget cuts – will not promote optimum levels of economic advancement and human development in rural areas. Investment is required so that the state can discharge its core regulatory and public good responsibilities as well as create the appropriate 'enabling environment' for markets. In addition there is the concern with distributional issues, the 'social efficiency' alluded to above, the achievement of which often requires additional state intervention. Certainly most of these expenditures will have to be financed domestically, but donors can play an important role in helping to develop the public sector capacity which is required to manage reform, and in supporting the type of innovative partnerships which are described in Chapter 5. Successful partnerships can then stand as examples for reform elsewhere, more effectively so if donors support horizontal learning networks and the rapid dissemination of lessons learned.

If such a vision is to be realised, however, there will have to be greater consideration given to identifying which countries should be receiving aid financing (which have demonstrated commitment to the notion of reform and a willingness to innovate?), and greater control exercised over the quality of aid interventions. Increased donor coordination will be important so that the lessons learned by one donor (and the country with which it is working) are quickly passed on to others. Perhaps most important, however, is the change that will be required in the overall 'direction' of aid. Thus far, most directly disbursed donor spending has gone to or through counterpart ministries or government offices in recipient countries. If the pluralistic and institutionally diverse view of development presented in this book is accurate, then this may no longer be appropriate. Increased funding of autonomous or semi-autonomous development funds – neutral forums through which various parties can engage with each other and find the most efficient solutions to the problems faced by rural people – is likely to be desirable. We would fully expect that many of these new solutions to old rural problems will involve multi-agency partnerships between new and reforming institutions both within and outside the public sectors.

# NOTES

## 1 Introduction

1 Given the severity of market imperfections, *a priori* theoretical analysis can be adduced in equal measure to support government and private sector provision. There are few theoretical reasons why one should be better than the other (Colclough, 1991).
2 These are issues which the private sector is prone to neglect, or has no incentive to consider.
3 Market failure is essentially a theoretical concept, though based on real observations. Government failure is an empirical concept. Proponents of this view triumphed the empirical over the theoretical or potential.

## 3 Reforming the state

1 Eco-tourism has been the boom industry of Costa Rica and is now the main earner of foreign exchange. Therefore, economic and environmental concerns are closely intertwined.
2 The Ministry of Forestry remains responsible for advisory and regulatory functions. The Department of Conservation has taken over conservation issues and a newly created Forestry Corporation is responsible for commercial forestry activities.
3 See the section in Chapter 4 on the private commercial sector for more detailed discussion of the economic concepts underlying reform.

## 4 Non-state approaches

1 It should be noted that the ODI research programme from which this book is drawn has not addressed all these areas in equal detail; this will be apparent in the text.
2 Current convention denotes resources as 'common pool' while management structures/systems are known as 'common property' regimes.
3 For their part, the unions view the researchers' approach as overly academic and long term in orientation.
4 Holt-Gimenez (1996) notes that after more than a decade of farmer-to-farmer extension in Nicaragua, still fewer than a dozen out of several hundred public sector researchers are prepared to support the farmer promoters, who by now total more than 300.

5   Donors are developing an increasingly critical attitude to NGOs; between 1988 and 1995, at least eleven official donors initiated assessments of the impact of NGO developmental activities partly funded by them (ODI, 1996).

6   Problems centred round the trading of vouchers and, more particularly, the fact that farmers were colluding with the supposed extension providers so that vouchers were handed over without the provision of any services. The 'providers' then redeemed the vouchers with the government and split the proceeds in some proportion with the farmers.

7   One reason for this high proportion of private sector research might lie in the fact that LDCs effectively do not do basic research – the type of research which is rarely attractive to private companies – and hence a greater proportion of what research is done might be 'eligible' for private sector supply.

## 5 The interface

1   Organised by the International Institute for Rural Reconstruction, ODI and World Neighbors in the Philippines in July 1995 (see Scarborough *et al.* (eds), forthcoming).

2   This is what has happened in the successful CODEFORSA/Ministry of Natural Resources, Energy and Mines (MIRENEM) partnership in Costa Rica (Richards *et al.*, 1996).

# BIBLIOGRAPHY

Agarwal, A. and Narain, S. (1989) *Towards Green Villages*, Delhi: Centre for Science and Environment.

Alsop, R.G. (1998) 'A Donor's Perspective and Experience of Process and Process Monitoring', in Mosse, D., Farrington, J. and Rew, A. (eds) *Development as Process: Concepts and Methods for Working with Complexity*, London: Routledge.

Alsop, R.G., Farrington, J., Gilbert, E.H. and Khandelwal, R. (forthcoming) *Multi-agency Approaches to Agricultural Technology Generation and Transfer in India*, (to be submitted to Sage).

Alsop, R.G., Khandelwal, R., Gilbert, E.H. and Farrington, J. (1996) 'The Human Capital Dimension of Collaboration among Government, NGOs and Farm Families: Comparative Advantage, Complications, and Observations from an Indian Case', *Agriculture and Human Values* 13 (12), pp. 3–12.

Anderson, J.R. (ed.) (1994) *Agricultural Technology: Policy Issues for the International Community*, Wallingford: CAB International.

Anderson, J.R. and de Haan, C. (eds) (1992) *Public and Private Roles in Agricultural Development: Proceedings of the Twelfth Agricultural Sector Symposium*, Washington, DC: World Bank Publications.

Antholt, C.H. (1994) 'Getting Ready for the Twenty-First Century: Technical Change and Institutional Modernisation in Agriculture', World Bank Technical Paper No. 217, Washington, DC: World Bank Publications.

Arnaiz, M.E.O., Merrill-Sands, D.M. and Mukwende, B. (1995) 'The Zimbabwe Farmers' Union: Its Current and Potential Role in Technology Development and Transfer', *Agricultural Research and Extension Network*, London: Overseas Development Institute.

Baile, P. (1996) 'Farmer-to-farmer Extension: Pedro Baile's Experience', Agricultural Research and Extension Network Paper 59a, pp. 45–8, London: Overseas Development Institute.

Bardhan, P. (1997) *The Role of Governance in Economic Development: A Political Economy Approach*, Paris: OECD.

Bates, R.H. (1981) *Markets and States in Tropical Africa: The Political Basis of Agricultural Policies*, Berkeley: University of California Press.

Bates, R.H. (1989) 'The Reality of Structural Adjustment: A Sceptical Appraisal', in S. Commander (ed.) *Structural Adjustment and Agriculture: Theory and Practice in Africa and Latin America*, London: Overseas Development Institute.

Batley, R. (1996) 'Privatisation of Services: Can Government Manage it?' *Development Research Insights*, London and Brighton: Overseas Development Institute and Institute of Development Studies.

Bebbington, A.J. (1997) 'Crises and Transition: Non-Governmental Organisations and Political Economic Change in the Andean Region', Agricultural Research and Extension Network Paper No. 76, London: Overseas Development Institute.

Bebbington, A. and Sotomayor, O. (1995) *Demand-led and Poverty-oriented ... or just Sub-contracted and Efficient? Learning from (Semi-)Privatised Technology Transfer Programmes in Chile*, Oxford: Food Studies Group.

Bebbington, A.J., Quisbert, J. and Trujillo, G. (1996) 'Technology and Rural Development Strategies in a Base Economic Organisation: El Ceibo Ltd. Federation of Agricultural Cooperatives', Agricultural Research and Extension Network Paper No. 62, London: Overseas Development Institute.

Behnke, R. (1995) *The Administration of Collective Institutions for Pastoral Resource Management*, London: Overseas Development Institute.

Behnke, R. and Kerven, C. (1994) 'Redesigning for Risk: Tracking and Buffering Environmental Variability in Africa's Rangelands', Natural Resource Perspectives Paper No. 1, London: Overseas Development Institute.

Beynon, J. and Duncan, A. (1996) *Financing of Agricultural Research and Extension for Smallholder Farmers in Sub-Saharan Africa: Summary Report and Guidelines*, Oxford: Food Studies Group.

Beynon, J., Duncan, A. and Jones, S. (1995) *Financing of Agricultural Research and Extension for Smallholder Farmers in Sub-Saharan Africa: Literature Review*, Oxford: Food Studies Group.

Beynon, J. and Mbogoh, S. (1996) *Financing Agricultural Research and Extension for Smallholder Farmers in Sub-Saharan Africa: The Case of Agricultural Research in Kenya*, Oxford: Food Studies Group.

Bingen, J., Carney, D. and Dembele, E. (1995) 'The Malian Union of Cotton and Food Crop Producers: Its Current and Potential Role in Technology Development and Transfer', Agricultural Research and Extension Network, London: Overseas Development Institute.

Brenner, C. (1991) *Biotechnology and Developing Country Agriculture: The Case of Maize*, Paris: OECD.

Bromley, D.W. (1989) *Economic Interests and Institutions: The Conceptual Foundations of Public Policy*, New York and Oxford: Basil Blackwell.

Bunch, R. (1996) 'People-centred Agricultural Development: Principles of Extension for Achieving Long-term Impact', Agricultural Research and Extension Network Paper No. 59a, pp. 11–18, London: Overseas Development Institute.

Burnside, C. and Dollar, D. (1996) *Aid Policies and Growth*, (first draft) November 1996

Carney, D. (1995a) 'Changing Public and Private Roles in Agricultural Service Provision: A Literature Survey', Working Paper No. 81, London: Overseas Development Institute.

Carney, D. (1995b) 'Management and Supply in Agriculture and Natural Resources: Is Decentralisation the Answer?', Natural Resource Perspectives Paper No. 4, London: Overseas Development Institute.

Carney, D. (1995c) 'The Changing Public Role in Services to Agriculture: A Framework for Analysis', Food Policy 20 (6), pp. 521–8.

Carney, D. (1996a) 'Formal Farmers' Organisations in the Agricultural Technology System: Current Roles and Future Challenges', Natural Resource Perspectives Paper No. 14, London: Overseas Development Institute.

Carney, D. (1996b) Farmers' Organisations in South Africa, Johannesburg: LAPC.

Carney, D. (1997) The Role of ODA in Agricultural Development, London: Overseas Development Institute.

CECAT/RCRE (1996) Research Study on Farmers' Specialised Technical Associations in China, Beijing: CECAT.

Chowdhury, M.K. and Gilbert, E.H. (1996) 'Reforming Agricultural Extension in Bangladesh: Blending Greater Participation and Sustainability with Institutional Strengthening', Agricultural Research and Extension Network Paper No. 61, London: Overseas Development Institute.

Colchester, M. (1994) 'Sustaining the Forests: The Community-based Approach in South and South East Asia', Development and Change 25 (1), pp. 69–100.

Colclough, C. (1991) 'Structuralism versus Neo-liberalism: An Introduction', in Colclough, C. and Manor, J. (eds), States or Markets? Neo-liberalism and the Development Policy Debate, Oxford: Clarendon Press.

Colclough, C. and Manor, J. (eds) (1991) States or Markets? Neo-liberalism and the Development Policy Debate, Oxford: Clarendon Press.

Collion, M-H. (1996) Suivi du Partenariat Recherche/Vulgarisation/Organisations de Producteurs et Implications pour les Systèmes de Recherche et Vulgarisation, mimeo, Washington DC: World Bank.

Commander, S. (ed.) (1989) Structural Adjustment and Agriculture: Theory and Practice in Africa and Latin America, London: Overseas Development Institute.

Cox, J. and Behnke, R. (1995) The Control of Communal Resources in Semi-Arid Namibia: Current Trends and Policy Issues, London: Overseas Development Institute.

Cromwell, E. (1992) 'The Impact of Economic Reform on the Performance of the Seed Sector in Eastern and Southern Africa', OECD Development Centre Technical Paper No. 68, Paris: OECD.

Crook, R. and Manor, J. (1994) 'Enhancing Participation and Institutional Performance: Democratic Decentralisation in South Asia and West Africa', Report to Overseas Development Administration, Brighton: Institute of Development Studies.

Dalrymple, D.G. and Srivastava, J.P. (1994) 'Transfer of Plant Cultivars: Seeds, Sectors and Society', in Anderson, J.R. (ed.), Agricultural Technology: Policy Issues for the International Community, Wallingford: CAB International.

Davis, S. and Wali, A. (1993) 'Indigenous Territories and Tropical Forest Management in Latin America', Policy Research Working Paper Series 1100, Environmental Assessments and Programs Division, World Bank, Washington, DC: World Bank Publications.

Echeverria, R.G., Trigo, E.J. and Byerlee, D. (1996) *Institutional Change and Effective Financing of Agricultural Research in Latin America*, World Bank Technical Paper No. 330, Washington, DC: World Bank.

Eyzaguirre, P. (1996) *Agriculture and Environmental Research in Small Countries: Innovative Approaches to Strategic Planning*, Chichester: Wiley.

FAO (1996) 'Food Production: The Critical Role of Water', Technical background document no. 7, prepared for the World Food Summit, November 1996.

Farrington, J. (1994) 'Public Sector Agricultural Extension: Is There Life after Structural Adjustment?' Natural Resource Perspectives Paper No. 2, London: Overseas Development Institute.

Farrington, J. and Bebbington, A.J. (1993) *Reluctant Partners? Non-Governmental Organisations, the State and Sustainable Agricultural Development*, London: Routledge.

Farrington, J. and Lewis, D.J. (eds) (1993) *Non-Governmental Organisation and the State in Asia*, London: Routledge.

Garcia, M. (1994) 'Malnutrition and Food Insecurity Projections, 2020', 2020 Project Brief No. 6, Washington, DC: IFPRI.

Garg, S. (1995) 'Experimenting with Farmer Led Extension Strategies in Rajasthan', Paper submitted to International Workshop on Farmer-led Approaches to Agricultural Extension held at IIRR, Silang, Cavite, Philippines on 17–22 July 1995.

Gordon, D.F. (1996) 'Sustaining Economic Reform Under Political Liberalisation in Africa: Issues and Implications', *World Development* 24 (9), pp. 1527–37.

Grindle, M.S. and Hilderbrand, M.E. (1995) 'Building Sustainable Capacity in the Public Sector: What can be done?' *Public Administration and Development* 15, pp. 441–63.

Grobman, A. (1992) 'Fostering a Fledgling Seed Industry', in Anderson, J.R. and de Haan, C. (eds), *Public and Private Roles in Agricultural Development: Proceedings of the Twelfth Agricultural Sector Symposium*, Washington, DC: World Bank Publications.

Gros, J.-G. (1994) 'Of Cattle, Farmers, Veterinarians and the World Bank: the Political Economy of Veterinary Service Privatisation in Cameroun', *Public Administration and Development* 14 (1), pp. 37–51.

Hardin, G. (1968) 'The Tragedy of the Commons'. *Science* 162, pp. 1243–8.

Hashemi, S.M., Schuler, S.R. and Riley, A.P. (1996) 'Rural Credit Programmes and Women's Empowerment in Bangladesh', *World Development* 24 (4), pp. 635–53.

Hobley, M. (1995) *Institutional Change within the Forestry Sector in South Asia: Centralised Decentralisation*, London: Overseas Development Institute.

Hobley, M. and Shah, K. (1996a) *What makes a Local Organisation Robust? Collective Resource Management in India and Nepal with Particular Reference to the Haryana Shavliks*, London: Overseas Development Institute.

Hobley, M. and Shah, K. (1996b) 'What makes a Local Organisation Robust? Evidence from India and Nepal', Natural Resource Perspectives Paper No. 11, London: Overseas Development Institute.

Holt-Gimenez, E. (1996) 'The Campesino-a-Campesino Movement: Farmer-led Agricultural Extension', Agricultural Research and Extension Network Paper 59a, pp. 1–10, London: Overseas Development Institute.

Husain, I. and Faruqee, R. (eds) (1994) *Adjustment in Africa: Lessons from Country Case Studies*, Washington, DC: World Bank Publications.

IFPRI. (1995) *A 2020 Vision for Food, Agriculture and the Environment – The Vision, Challenge and Recommended Action*, October 1995, Washington DC: IFPRI.

Jaeger, W.K. (1992) 'The Effects of Economic Policies on African Agriculture', World Bank Discussion Paper 147, Washington, DC: World Bank Publications.

Kaimowitz, D. (ed.) (1990) *Making the Link: Agricultural Research and Technology Transfer in Developing Countries*, Boulder, Colorado: Westview.

Kerven, C. and Cox, J. (1996) *Changing Rangeland Management on the Borana Plateau, Ethiopia*, London: Overseas Development Institute.

Killick, T. (1997) 'What Future for Aid?', paper prepared for Fourth Expert Group Meeting, Santiago, 1–10 January 1997.

Kingsley, M.A. and Musante, P. (1996) 'Activities for Developing Linkages and Cooperative Exchanges among Farmers' Organisations, NGOs, GOs and Researchers', Agricultural Research and Extension Network Paper No. 59b, pp. 5–13, London: Overseas Development Institute.

Kirkpatrick, C. (1996) 'Privatisation (An Introduction)', *Development Research Insights*, London and Brighton: Overseas Development Institute and Institute of Development Studies.

Klitgaard, R. (1991) *Adjusting to Reality: Beyond 'State Versus Market' in Economic Development*, San Francisco: ICS Press.

Lewa, P.M. (1995) 'Kenya's Cereal Sector Reform Programme: Managing the Politics of Reform', paper presented at University of Birmingham workshop on *The Changing Public Role in Services to Agriculture*, 24 April 1995.

Mans, D. (1994) 'Tanzania: Resolute Action', in Husain, I. and Faruqee, R. (eds), *Adjustment in Africa: Lessons from Country Case Studies*, Washington, DC: World Bank Publications.

McKean, M. (1995) 'Common Property: What is it, What is it Good for, and What makes it Work?', paper presented at the International Conference on 'Chinese Rural Collectives and Voluntary Organisations: Between State Organisation and Private Interest', Sinological Institute, University of Leiden, 9–13 January 1995.

Mehta, L. (1996) *Kutch, the Sardar Sarovar Project and the Socio-Economic Component in Water Resources Management*, London: Overseas Development Institute.

Merrill-Sands, D. and Collion, M-H. (1993) 'Making the Farmer's Voice Count in Agricultural Research', *Quarterly Journal of International Agriculture* 32 (3), pp. 260–79.

Merrill-Sands, D. and Kaimowitz, D. (1991) *The Technology Triangle: Linking Farmers Technology Transfer Agents and Agricultural Researchers*, The Hague: ISNAR.

Miclat-Teves, A.G. and Lewis, D. (1993) 'NGO–Government Interaction in the Philippines: Overview', in Farrington, J. and Lewis, D. (eds), *Non-Governmental Organisations and the State in Asia*, London: Routledge.

Mosse, D. (1994) 'Authority, Gender and Knowledge: Theoretical Reflections on the Practice of Participatory Rural Appraisal', Agricultural Research and Extension Network Paper No. 44, London: Overseas Development Institute.

Mosse, D. (with the KRIBP team) (1996) 'Local Institutions and Farming Systems Development: Thoughts from a Project in Tribal Western India', Agricultural Research and Extension Network Paper No. 64, London: Overseas Development Institute.

Mosse, D., Farrington, J. and Rew, A. (eds) (forthcoming) *Development as Process: Concepts and Methods for Working with Complexity*, London: Routledge.

ODI (1996) 'The Impact of NGO Development Project', Briefing Paper 1996 (2), London: Overseas Development Institute.

OECD/DAC (1995) *Support of Private Sector Development*, Paris: OECD.

OECD/DAC (1997) *Development Projects as Policy Experiments*, London: Routledge.

Okali, C., Sumberg, J. and Farrington, J. (1994) *Farmer Participatory Research*, London: Intermediate Technology.

Ostrom, E. (1990) *Governing the Commons: the Evolution of Institutions for Collective Action*, Cambridge: Cambridge University Press.

Ostrom, E. (1992) *Crafting Institutions for Self-Governing Irrigation Systems*, San Francisco: Institute for Contemporary Studies Press.

Pandit, B.H. (1996) 'The Nepal Agroforestry Foundation's Approach to Farmer-led Extension', Agricultural Research and Extension Network Paper No. 59a, pp. 28–35, London: Overseas Development Institute.

Pant, M. and Cox, J. (1996) *Common Property – Common Good? Catchment Level Water Management in Nepal*, London: Overseas Development Institute.

Pardey, P.G., Roseboom, J. and Anderson, J.R. (eds) (1991) *Agricultural Research Policy: International Quantitative Perspectives*, Cambridge: Cambridge University Press.

Pray, C.E. and Echeverria, R.G. (1991) 'Private Sector Agricultural Research in Less Developed Countries', in Pardey, P.G., Roseboom, J. and Anderson, J.R. (eds), *Agricultural Research Policy: International Quantitative Perspectives*, Cambridge: Cambridge University Press.

Ramirez, J. (1994) 'Irrigation Investment and Management Transfer in Colombia', paper prepared for International Conference on Irrigation Management Transfer, Wuhan, China, September 1994.

Ranis, G. (1996) 'Successes and Failures of Development Experience Since the 1980s', main paper presented at the Development Thinking and Practice Conference, Washington, DC, 3–5 September 1996.

Ravnborg, H.M. and Ashby, J.A. (1996) 'Organising for Local Level Watershed Management: Lessons from the Rio Cabuyal Watershed, Colombia', Agricultural Research and Extension Network Paper No. 65, London: Overseas Development Institute.

Richards, M. (1995) *A Review of Recent Institutional Change in the Forest Sector of Latin America*, London: Overseas Development Institute.

Richards, M. (1996) 'Stabilising the Amazon Frontier: Technology, Institutions and Policies', Natural Resource Perspectives Paper No. 10, London: Overseas Development Institute.

Richards, M. (1997a) 'Missing a Moving Target? Colonist Technological Development on the Amazon Frontier', ODI Research Study, London: ODI.

Richards, M. (1997b) 'Common Property Resource Institutions and Forest Management in Latin America?', *Development and Change* 28 (1), pp. 95–117.

Richards, M., Navarro, G., Vargas, A. and Davies, J. (1996) 'Decentralisation and the Promotion of Sustainable Forest Management and Conservation in Central America', ODI Working Paper No. 93, London: ODI.

Röling, N. (1990) 'The Agricultural Research–Technology Transfer Interface: A Knowledge Systems Perspective,' in Kaimowitz, D. (ed.), *Making the Link: Agricultural Research and Technology Transfer in Developing Countries*, Boulder, Colorado: Westview.

Rondinelli, D. (1993) *Evaluation of Programs Promoting Participatory Development and Good Governance: Synthesis Report*, Paris: OECD.

Samad, M., Dingle, M.A. and Shafique, M.S. (1994) 'Political and Economic Dimensions of Privatisation and Turnover: Irrigation Schemes in Sudan', paper prepared for International Conference on Irrigation Management Transfer, Wuhan, China, September 1994.

Sarin, M. (1996) 'Actions of the Voiceless: the Challenge of Addressing Subterranean Conflicts Related to Marginalised Groups and Women in Communities', FAO Email conference on Addressing Natural Resource Conflicts through Community Forestry.

Scarborough, V. (1996) 'Philippines Workshop Compilation: Editorial Introduction', Agricultural Research and Extension Network Papers 59a–c, London: Overseas Development Institute.

Scarborough, V., Killough, S., Johnson, D. and Farrington, J. (eds) (forthcoming) *Farmer-led Extension: Concepts and Practices*, London: Intermediate Technology Publications.

Schiff, M. and Valdés, A. (1992) *The Plundering of Agriculture in Developing Countries*, Washington, DC: World Bank Publications.

Schwartz, L.A. (1994) 'The Role of the Private Sector in Agricultural Research and Extension: Economic Analyses and Case Studies', Agricultural Research and Extension Network Paper No. 48, London: Overseas Development Institute.

Sen, A. (1996) 'Development Thinking at the Beginning of the 21st Century', main paper presented at the Development Thinking and Practice Conference, Washington DC, 3–5 September 1996.

Shah, A. (1995) 'NGO–GO Interactions in Watershed Development: Experiences from Gujarat (India)', Agricultural Research and Extension Network Paper No. 56, London: Overseas Development Institute.

Shikavoti, G.P. (1994) 'Management Transfer of Agency-Managed Irrigation Systems in Nepal', paper prepared for International Conference on Irrigation Management Transfer, Wuhan, China, September 1994.

SIDA (1996) *Aid Dependency: Causes, Symptoms and Remedies*, Stockholm: SIDA.

Sims, H. and Leonard, D. (1990) 'The Political Economy of the Development and Transfer of Agricultural Technologies Perspective' in Kaimowitz, D. (ed.), *Making the Link: Agricultural Research and Technology Transfer in Developing Countries*, Boulder, Colorado: Westview.

Sinaga, N. and Wodicka, S. (1996) 'Farmer-based Extension in the Marginal Uplands of Sumba, Indonesia: A Case Study of Tananua Experience', Agricultural

Research and Extension Network Paper No. 59a, pp. 19–27, London: Overseas Development Institute.

Southgate, D. and Runge, C.F. (1990) 'The Institutional Origins of Deforestation in Latin America', paper prepared for Conference on Economic Catalysts to Ecological Change, University of Florida Centre for Latin American Studies, February 1990.

Sperling, L., Scheidegger, U. and Buruchara, R. (1996) 'Designing Seed Systems with Small Farmers: Principles Derived from Bean Research in the Great Lakes Region of Africa', Agricultural Research and Extension Network Paper No. 60, London: Overseas Development Institute.

Srivastava, J.P. and Jaffee, S. (1993) 'Best Practices for Moving Seed Technology: New Approaches to Doing Business', World Bank Technical Paper No. 213, Washington, DC: World Bank Publications.

Streeten, P. (1996) 'Globalisation and Competitiveness: What are the implications for Development Thinking and Practice?', main paper presented at the Development Thinking and Practice Conference, Washington, DC, 3–5 September 1996.

Stringfellow, R. and McKone, C. (1996) *The Provision of Agricultural Services through Self-Help in Sub-Saharan Africa: Zimbabwe Case Study*, Chatham: NRI/Plunkett Foundation.

Stringfellow, R., Lucey, T. and McKone, C. (1996) *The Provision of Agricultural Services through Self-Help in Sub-Saharan Africa: Uganda Case Study*, Chatham: NRI/Plunkett Foundation.

Stringfellow, R., Coulter, J., Lucey, T., McKone, C. and Hussain, A. (1997) 'Improving the Access of Smallholder to Agricultural Services in Sub-Saharan Africa: Farmer Cooperation and the Role of the Donor Community', Natural Resources Perspectives Paper No. 20, London: Overseas Development Institute.

Swamy, G. (1994) 'Kenya: Patchy, Intermittent Commitment', in I. Husain and R. Faruquee (eds) *Adjustment in Africa: Lessons from Country Case Studies*, Washington, DC: World Bank Publications.

Timmer, C.P. (ed.) (1991) *Agriculture and the State*, Ithaca: Cornell University Press.

Toye, J. (1991) 'Is There a New Political Economy of Development?', in Colclough, C. and Manor, J. (eds), *States or Markets? Neo-liberalism and the Development Policy Debate*, Oxford: Clarendon Press.

Tripp, R. and Gisselquist, D. (1996) 'A Fresh Look at Agricultural Input Regulation', Natural Resource Perspectives No. 9, London: Overseas Development Institute.

Turral, H. (1995a) 'Recent Trends in Irrigation Management – Changing Directions for the Public Sector', Natural Resource Perspectives No. 5, London: Overseas Development Institute.

Turral, H. (1995b) 'Devolution of Management in Public Irrigation Systems: Cost Shedding, Empowerment and Performance: A Review', Working Paper No. 80, London: Overseas Development Institute.

Turral, H. (1995c) *Case Studies of Urban/Rural Competition for Water: A Synthesis*, London: Overseas Development Institute.

Umali, D., Feder, G. and de Haan, C. (1992) 'The Balance between Public and Private Sector Activities in the Delivery of Livestock Services', World Bank Discussion Paper No. 163, Washington, DC: World Bank Publications.

Utting, P. (1993) *Trees, People and Power: Social Dimensions of Deforestation and Forest Protection in Central America*, London: Earthscan.

Veblen, T. (1978) 'Forest preservation in the Western Highlands of Guatemala', *Geographical Review* 68, pp. 417–34; cited in Utting (1993).

Vickers, J. and Yarrow, G. (1988) *Privatisation: An Economic Analysis*, Cambridge, MA: MIT Press.

Wade, R. (1988) *Village Republics: Economic Conditions for Collective Action in South India*, Cambridge: Cambridge University Press.

Walker, M. (1996) 'Strengthening Agricultural Extension in Bangladesh: A Short Update Following from Network Papers 59c and 61', *Agricultural Research and Extension Network Newsletter* 34, pp. 13–15, London: Overseas Development Institute.

World Bank (1994) *Adjustment in Africa: Reforms, Results and the Road Ahead*, New York: Oxford University Press.

World Bank (Africa Region) (1995) *Continent in Transition: Sub-Saharan Africa in the Mid-1990s*, Washington, DC: World Bank.

World Bank (1997) *The State in a Changing World*, New York: Oxford University Press.

# INDEX

access, open 14, 17, 28, 38, 90, 96
accountability 5, 21, 51, 61, 77, 86, 89, 92, 94, 98, 101
accounts 41, 68
adjustment, structural 15, 63
advice 6, 33, 47–8, 65, 93
advocacy 61–2
Africa 17–20, 25, 34, 63–5; *see also individual countries*
Agarwal, A. 12
agriculture 1, 2, 6, 7, 9–12, 18, 28, 30, 35, 47, 50–5 *passim*, 57, 59, 63–9, 75–84, 90–4 *passim*; Development Project (ADP) 80, 81; Research Fund (ARF) 80, 83
aid 2, 28, 34, 55, 85, 103–4
Alsop, R.G. 33, 76, 78, 83
Amazonia 14, 50
America, Central 15, 47, 57, 59; Latin 14, 23, 25, 43, 47, 55, 66; *see also individual countries*
Antholt, C.H. 9, 11, 12
Ashby, J.A. 40, 56, 71
Asia 15, 25, 57, 63; east 1; south 14, 74, 90, 95; *see also individual countries*
Australia 19, 96

Baile, P. 59
bananas 29
Bangladesh 10–11, 16, 22–4, 27, 29–34 *passim*, 36, 56, 57, 65; Agricultural Technical Committees 24; BRAC 56; Grameen Bank 57
banks 43, 63
Bardhan, P. 22
bark stripping 13

Bates, R.H. 10, 68
Batley, R. 24
Bebbington, A. 54–6, 60–3 *passim*, 69
Behnke, R. 17–20 *passim*, 30, 38, 39, 42
Beynon, J. 34, 63–5 *passim*
bhabbar grass 28, 43
Bingen, J. 48, 51, 54
biodiversity 14, 23
Bolivia 48, 54; Coraca-Potosi 54; El Ceibo cocoa cooperatives 47–8, 54, 62, 73
Boran/Borana 18, 20
boundaries 17, 38, 39, 44
Brazil 14, 23, 36, 50, 69, 73
Brenner, C. 64, 69
Bromley, D.W. 38
Bunch, R. 58, 59
Burnside, C. 103

Cameroon 68
capacity 29–33 *passim*, 41, 53; building 30, 32–3, 95, 100
Caribbean 25
Carney, D. 4, 12, 22, 30, 34, 37, 49–51 *passim*, 53–4, 63, 67, 69, 103
CECAT/RCRE 32
Chile 63
China 1, 15, 31–2, 35, 62, 65, 66
Chowdhury, M.K. 11, 22, 27, 75
civil service 1, 6, 7, 21–2, 26, 31–4, 77, 83, 86, 89, 91, 92, 99, 101
climate 23, 66
cocoa 47–8, 54
Colchester, M. 12
Colclough, C. 1, 7, 33
Collion, M-H. 9

Colombia 67
common pool resources 20, 37–47, 55, 72, 73, 85, 89
companies, commercial 9, 12, 14, 16, 62–72, 89; multinational 37, 68, 72
competition 4, 23, 65, 69, 71
conflict resolution 41, 46, 91, 92, 94, 96
conservation 14, 20, 29, 33, 58–9
contracting 3, 27, 62, 71, 76–7; sub- 62
cooperatives 47, 54, 68, 69
coordination 24, 25, 28–30, 36, 89; donor 104
corruption 2, 9, 10, 13, 14, 28, 32, 54, 60, 69
Costa Rica 22, 24, 26, 29, 30, 60, 75; CODEFORSA 60; Tortugero Conservation area 29
costs 9, 15, 22, 34, 36, 66–7; recovery 3, 15, 69, 95; recurrent 9, 11, 54; reduction 22, 34, 67; sharing 11
cotton 48, 50, 54
Cox, J. 16–18 passim, 20, 30, 39, 41, 55
credit 11, 19, 31–3, 42–3, 48, 56, 63, 67, 69, 76, 80, 93
Cromwell, E. 69
Crook, R. 32

Dalrymple, D.G. 68
DANIDA 48, 51, 54
decentralisation 2, 3, 12, 15, 22–4, 26–33 passim, 36, 89, 100
degradation, land 18, 20
democratisation 14, 55, 92
demonstrations 31, 76, 79
deregulation 2, 4, 65, 68
Dollar, D. 103
donors 5, 6, 9, 15, 22, 27, 28, 34, 36, 53, 54, 75, 91–6 passim, 103–4
Duncan, A. 63, 65

Echeverria, R.G. 65, 68
education 5, 31–2, 34, 47, 57, 71
effectiveness 5–6, 35, 72, 88, 98
efficiency 5–6, 12, 22–5, 72, 88, 98, 101; social 5, 6, 35, 70, 99, 104
ejidos 39, 42
elites/elitism 7, 10, 12, 17, 32, 69, 90, 96, 100
employment 4, 22, 34, 61
enclosure, land 18, 30, 96

environment, enabling 28, 42–3, 70–1, 104
environmental factors 4, 23, 24, 32, 36, 75
equity 4, 6, 14–16 passim, 41, 44, 92
erosion, soil 23, 40
Ethiopia 17–20 passim, 41
European Union 22, 68–9
expenditure, public 4, 11, 22–3, 25, 33–4
extension 9–12, 22–36 passim, 48, 49, 56, 59, 60, 63, 65, 69, 70, 76, 79–81, 92–3, 100; farmer-led 59–60; farmer-to-farmer 47, 55, 57–60, 73; para-extension workers (PEWs) 80, 81, 93; village-level (VLWs) 76, 92–3
Eyzaguirre, P. 25

Farrington, J. 56, 60, 69
felling, tree 13, 14, 23, 30, 40
fencing 17, 18, 96
fertiliser 63, 65, 66, 76
financing 22, 26, 27, 33–4, 42, 54, 67–8, 85, 91, 92; see also aid
flexibility 19, 21, 45, 60, 75, 83, 91, 95, 96
Ford Foundation 75–80 passim
foreign exchange 68
forest departments 12, 14, 15, 26, 28, 29, 33, 39, 91, 95
forest products, non-timber 95–6
forestry 2, 6, 9, 12–15, 24, 28, 30, 38, 39, 44–6, 58–9, 63, 66–7, 75, 87, 95–6; joint forest management (JFM) 14, 26, 39–41 passim, 77, 84, 87, 90, 95; village committees 39
fruit 48–9

Gabbra 18
Garcia, M. 7
Garg, S. 75
GATT/WTO 69
Gilbert, E.H. 11, 22, 27, 75
Gisselquist, D. 23
government departments 3, 21, 22, 24, 37, 92; coordination 24, 28–30, 89
grazing 13, 86, 96
Green Revolution 11, 15
Grindle, M.S. 29, 30, 32, 35, 101
Gros, J-G. 68
group formation 10, 33, 40–1, 56–7,

61, 83, 85, 86, 98; CPR 38–47;
    farmer 33, 37, 47–51 *passim*, 62,
    69, 72, 75, 81, 84, 86; membership
    61, 73, 74, 90, 92, 102; pastoral 39;
    user 16, 22, 23, 26, 32, 36, 38, 39,
    42, 44–6, 70–1, 94
Guatemala 13
Gujarat 29, 61

Hardin, G. 17
Haryana 26, 28, 42, 43, 84
Hashemi, S.M. 57
health 5, 34, 56, 57, 65, 74
Hilderbrand, M.E. 29, 30, 32, 35, 101
Hobley, M. 15, 22, 26–31 *passim*, 34,
    41–6 *passim*, 75
Holt-Gimenez, E. 47, 58, 59
Honduras 23, 36
human resource development 6, 83

income 4, 19, 44, 61
India 11, 12, 14, 15, 28–30 *passim*, 39,
    41, 42, 57, 73, 75–86; *see also*
    *individual states*
Indonesia 12, 15, 24; Training
    Programme (IPM) 59
infrastructure 4, 15, 16, 20, 32, 66, 67,
    69, 71, 88, 89, 100, 103
innovation 11–12, 16, 60
inputs, agricultural 9–11 *passim*, 42,
    48–50 *passim*, 63, 65, 66, 76, 100
institutional factors 15, 16, 28, 30–1,
    85, 97–8, 100
intervention, state 5, 14, 23, 70, 89, 99,
    104
investment 15, 16, 32, 45, 63, 66, 68,
    69, 96, 100, 104
irrigation 9, 11, 15–16, 21, 24, 36, 39,
    40, 43, 44, 63, 67, 76, 85, 91, 94–5,
    100, 103; management transfer 16,
    26, 28, 32, 44, 91, 94
ITDG 59

Jaeger, W.K. 10
Jaffee, S. 65

Kaimowitz, D. 24, 29
Karnataka 32
Kenya 27, 64–6 *passim*; Breweries 64;
    Farmers' Union 48
Kerven, C. 17–20 *passim*, 41
Killick, T. 104

Kingsley, M.A. 59
Kirkpatrick, C. 4, 23
Klitgaard, R. 70
knowledge, farmer 47

labour 61
land 12, 13, 17–18, 42; reform 50;
    rights 13, 18, 96; tenure 13, 17–20
    *passim*, 39, 96, 100
landless 14, 23, 56
law, customary 18
learning 5, 7, 27, 78, 91, 97, 104
leases 28, 43
legal system 30, 42, 71
Leonard, D. 9, 49
Lewa, P.M. 27
Lewis, D. 57
licensing 13
linkages, extension-research 10, 11, 24,
    29–30, 36
loans, 2, 49, 52, 56, 63; repayment of
    42, 43, 49, 56–7
lobbying 50, 51, 61, 103
logging 14, 66
Lomé Convention 69

McKean, M. 43, 46
McKone, C. 51
mahogany 28
maintenance 9, 15, 91, 94, 97
maize 21, 64, 68, 76, 79
Mali Union of Cotton and Food Crop
    Producers (SYCOV) 48, 50–1, 54
Manor, J. 32
Mans, D. 21, 66
marketing 19, 21, 27, 37, 42, 48, 49,
    56, 63–5 *passim*, 93, 96; boards 9
markets 1, 5, 8, 16, 28, 43, 47, 63, 64,
    69–71 *passim*, 99; failure 7;
    liberalisation of 63, 64; parallel 1,
    62
mass media 93
Mbogoh, S. 64
Merrill-Sands, D. 9, 29
Mexico 14, 15, 23, 24, 36, 39, 42, 43,
    69
Miclat-Teves, A.G. 57
monitoring 23, 31, 35, 36, 46, 60, 78,
    79, 82, 92
monopolies 9, 23, 65, 71
moral hazard 71
Mosse, D. 28, 41, 57

Muchagata, Marcia 50
multi-agency approach 7–8, 10, 58, 87, 90–2, 99–100, 102; *see also* partnerships
Musante, P. 59

Namibia 17, 18, 30
Narain, S. 12
Nepal 16, 43, 44, 59, 73, 87; Agro-Forestry Foundation 58
New Zealand 29
NGOs 2, 7, 10, 11, 25, 32, 33, 37, 41–3, 47, 55–62, 69, 72–86, 89–92 *passim*, 95, 98–103 *passim*; and governments 55, 57, 60, 74–86, 99; weaknesses of 60–1

ODI 15, 57, 60, 78, 79
OECD/DAC 4, 30, 101
Okali, C. 9
organisations, farmer 7, 10, 25, 30, 47–55, 69, 72, 73, 85, 92; membership 2, 10, 37, 47–55, 74, 89, 90, 92, 95, 97, 100, 102
Ostrom, E. 16, 43, 46
ownership 3, 12, 21, 23, 26, 38, 41, 42, 71

Pakistan 15; Aga Khan Rural Support Programme 56
Pandit, B.H. 58, 59
Pant, M. 16, 39, 55
paradox, orthodox 32, 101; of power 101
parastatals 27, 37, 42, 71
participation, user 9, 11, 26, 38, 41–2, 51, 80, 86, 98
partnerships, state/non-profit organisations 74–87, 90–2, 99–100, 102–4 *passim*
pastoralism 2, 6, 17–20, 39, 63, 96; *see also* rangelands
patronage 16, 22, 26, 27, 31, 32, 67
pest control 50, 59, 67
Philippines 44, 57, 59
pilot schemes 44, 91, 94, 95
planning 23, 33, 41
political factors 7, 27–30, 32, 86, 101
ponds 20, 41
poor, rural 4, 5, 15, 34, 41, 55, 61, 69, 90, 99–101 *passim*

poverty 7, 42; reduction 5, 95, 99–101 *passim*, 104
Pray, C.E. 65
prices 9, 17, 50, 63, 68, 70
private sector 2–4 *passim*, 7, 24, 37, 62–72, 74, 75, 80, 88, 98, 102; *see also* companies
privatisation 1–3 *passim*, 14, 18, 19, 21, 23, 63, 71, 95
productivity 7, 11, 14, 15, 20, 44, 90
project approach 19–20
public sector 5–11, 62–72 *passim*, 75–86, 91–3, 96, 98–102 *passim*, 104; reform of 21–36, 88–9
punishment 45, 46

quangos 22

Rajasthan 7, 27, 33, 40, 75–84, 102
Ramirez, J. 67
ranches 14, 19, 96
rangelands 17–20, 30, 38, 39, 63, 96, 100; indigenous management of 18–20
Ranis, G. 70
Ravnborg, H.M. 40, 56, 71
reform 2–7, 20, 97–8; civil service 101; goals of 5–6, 98; institutional 15, 16; legal 42; state 3, 7, 21–36, 67, 88–9
regionalisation 25, 33, 36, 89
regulation 23, 24, 32, 39, 64, 71, 94, 102–3
rent seeking 16, 31, 32
research, agricultural 2, 9–12, 24–30 *passim*, 33, 36, 49, 50, 56, 59, 60, 63, 65, 68, 69, 80; ASARECA 25; CGIAR 25; CORAF 25; Global Forum on 25, 27, 36; Tocantins centre 50
resistance to change 26–8, 35, 77, 89
retraining 23, 32
Richards, M. 12, 14, 15, 22–4, 28–30 *passim*, 42, 43, 50, 55, 59, 60, 75
rights 42, 45; land 13, 18, 96; property 19, 71; user 12, 14, 39, 41, 94, 97; water 15, 17, 32, 94, 95
risk 27, 56, 65, 66
roads 66, 68
Roling, N. 9
Rondinelli, D. 101
rules 45–6

Samad, M. 67
sanctions 46, 90
Sarin, M. 40
savings 42, 43; and loans schemes 56–7
Scarborough, V. 12
Schiff, M. 10
Schwartz, L.A. 63
seeds 23, 37, 50, 56, 63, 64, 68, 69
Sen, A. 70
Shah, K. 15, 26, 27, 31, 34, 42–6
    passim, 56, 57, 61
Shikavoti, G.P. 43
SIDA 34, 104
Sims, H. 9, 49
Sinaga, N. 59
skills 6, 22, 24, 31–3 passim, 36, 41,
    51, 52, 62, 78, 83, 85, 91–3 passim
Sotomayor, S. 63
South Africa 18, 50
Sperling, L. 64
Sri Lanka 44
Srivasta, J.P. 65, 68
state, reform of 21–36, 67, 88–9; role
    1–5, 15, 70, 89, 93, 101–3 passim;
    withdrawal of 4, 16, 21, 33, 62, 88
Streeten, P. 70
strikes 48, 50, 81
Stringfellow, R. 49, 51, 52
subsidies 63, 67, 94
Sudan 67
Swamy, G. 66

Tamil Nadu 11, 16–17
Tanzania 21
taxes 48, 67, 70
technologies 9, 11, 16, 25, 30–3
    passim, 36, 47, 49–54, 57–60
    passim, 62, 76, 100
Thailand 12
Timmer, C.P. 70
titling, land 19
tobacco 65
Toye, J. 1
trade 1, 68, 69
'tragedy of the commons' 17
training 10, 11, 23, 31, 41–2, 48, 49,
    52, 54, 58, 80; and Visit 10–12, 31,
    69, 92
trials 31
Tripp, R. 23
Turkey 68

Turral, H. 15–17 passim, 27–8, 32,
    40–2 passim, 44, 63, 75
Twomlow, Steve 33

Udaipur 33, 75–84, 87; KVK 79, 82,
    83
Uganda 48–9; Farmers' Association 48,
    51, 54, 73
UK 5, 22, 103
Umali, D. 63
USA 19, 68, 96
users 7, 9, 14, 16–18, 21, 23, 33,
    38–40 passim, 45, 72, 97; see also
    rights
Uttar Pradesh 44
Utting, P. 13
UVAN Ltd 49

Valdes, A. 10
vanilla 49
Veblen, T. 13
veterinary services 20, 63, 68
Vickers, J. 71
voluntarism 59

Wade, R. 43, 46
Walker, M. 24
water resources 2, 6, 9, 15–17, 27–9,
    32, 38, 67, 94–5, 100; see also
    irrigation; Development Project
    (WDRP) 81
water user associations 39, 40, 42, 94
watersheds 55, 56, 61, 80, 94;
    Management Guidelines 77
welfare economics 4, 63–4
Wodicka, S. 59
women 10, 23, 34, 40, 52, 56
World Bank 2, 9, 10, 11, 22, 28, 63,
    77, 80, 81, 92, 98
World Development Report 35, 70

Yarrow, G. 71
yields 59, 68

Zambia 69
Zimbabwe 33, 48, 51, 59, 65;
    AGRITEX 59; Credit Board 51;
    Farmers Union 48, 51, 54, 59